Plastic Water

Plastic Water

The Social and Material Life of Bottled Water

Gay Hawkins, Emily Potter, and Kane Race

The MIT Press
Cambridge, Massachusetts
London, England

This book was set in Stone by the MIT Press. Printed on recycled paper and bound in the United States of America.

Library of Congress Cataloging-in-Publication Data

Hawkins, Gay.
Plastic water : the social and material life of bottled water / Gay Hawkins, Emily Potter, and Kane Race.
    pages   cm
Includes bibliographical references and index.
ISBN 978-0-262-02941-4 (hardcover : alk. paper)
1. Bottled water industry. 2. Bottled water industry—Social aspects. 3. Bottled water. 4. Bottled water—Social aspects. 5. Drinking water. 6. Drinking water—Social aspects. I. Potter, Emily. II. Race, Kane. III. Title.
HD9349.M542H39   2015
338.4'766361—dc23
2015001901

10   9   8   7   6   5   4   3   2   1

Contents

Acknowledgments

This book is the outcome of an Australian Research Council Discovery Grant, "From the Tap to the Bottle: The Social and Material Life of Bottled Water" (DP0879786). We thank this funding body for supporting this research: it would not have happened without it. Over the years of the study, we investigated bottles in action in Bangkok, Chennai, Évian, Hanoi, Tokyo, London, Australia, and the United States. This research was supported by numerous people, including Chippy Kurian and the Kurian family, Dr. Sekar Raghavan, Dr. Minh Duong, Pham Van Duc, Nguyen Thi Le Huyen, Le Huy Tuyen, Professor Hideshige Takada, and Susan and David Race. Numerous staff from major beverage companies were also immensely helpful in sharing their insights and perspectives and in giving us access to bottling plants and documents. They are not named in the text, but we appreciate their willingness to contribute. We also thank staff at the Polaris Institute for being available for interviews and providing us with an excellent overview of the development of their Inside the Bottle campaign. Our work was assisted by excellent research support from Ben Gook, Simon Sellars, Morgan Richards, Katy Green, Kirsty Leishman, Donald McKay, Rebecca Brown, and Georgie Isbister. Warwick Pearse provided specialist research work on the occupational and environmental impacts of the plastic villages in Hanoi and the wider impacts of PET recycling—this was especially valuable. We also thank Amanda McKittrick for the care she took sourcing and preparing images, and Sue Jarvis for being such an efficient and effective editor.

Over the course of the research, we were invited to present papers in various forums at the National University of Singapore, Johns Hopkins University, the University of Michigan, Goldsmiths, the University of Sydney, the University of Melbourne, the Royal Melbourne Institute of Technology, and the University of South Australia. We are enormously grateful to the audiences at these events for their feedback and challenging questions. Jane Bennett, David Halperin, Sarah Whatmore, Kay Anderson, Lesley Head, John Law, Damon Young, Marjorie Levinson, Lisa Disch, Noortje Marres, Robert Foster, Rosalyn Diprose, Tim Winter, Elizabeth Shove, Mathew Gandy, Mike Michael, Andrea Westermann, Jennifer Gabrys, Marsha Rosengarten, and Catherine Waldby were especially insightful in their responses and engagements, and the book is the better for it. We also thank the anonymous reviewers for the MIT Press, who gave excellent feedback on the original proposal and the final manuscript.

The book was written while the three authors worked in numerous workplaces. For the final three years, Gay Hawkins was at the Centre for Critical and Cultural Studies at the University of Queensland. This research center provided a superb environment for completing the study, and great thanks are owed for the extensive administrative support provided by Fergus Grealy and Rebecca Ralph. They kept everything on track with calm efficiency. Colleagues there also provided excellent and ongoing feedback on various stages of the manuscript. Thanks to Graeme Turner, Mark Andrejevic, Anna Pertierra, Melissa Bellanta, Anthea Taylor, Abi Loxham, Graeme Were, and Catherine Phillips for their invaluable contributions. Kane Race was in the Department of Gender and Cultural Studies at the University of Sydney during most of the study. This was an immensely stimulating and inspiring environment. Thanks to Catherine Driscoll, Elspeth Probyn, Melissa Gregg, Jane Park, and the students of Natures and Cultures of Bodies for their discussion of the project and their encouragement over the years. For much of the project, Emily Potter was a research fellow in the Centre for Citizenship and Globalisation at Deakin University, Melbourne, followed by her current position as senior lecturer in the School of Communication and Creative Arts, also at Deakin. These environments provided indispensable support and interest in the project. Special thanks go to Alison Huber for her sustaining intellectual and collegial generosity.

Finally, we thank all the people who have sent us links, information, photographs, and ideas during the course of our research. It is impossible to name them all, but two stand out. Early in the study, Michael Harris sent us the link to Lewis Black's anti–bottled water rant on YouTube. Cameron Tonkinwise sent us the link to the Brita company's FilterForGood "drinking oil" campaign. Both these gestures were wonderfully generous and provocative for our thinking.

Writing a book with three authors is challenging: while it is great to extend the reach of the research, it is also harder to manage the massive amount of data and information and to keep connected and focused across three cities. The great benefit is that we have had the opportunity to research and write in a deeply collaborative way, to test our ideas and arguments in a rigorous and generous environment, and to find a truly collective voice.

Earlier versions of the research have been published previously, but this work has been substantially extended and revised for the book:

Hawkins, G. The politics of bottled water. *Journal of Cultural Economy* 2 (1–2) (2009): 183–195.

Hawkins, G. Making water into a political material: The case of PET bottles. *Environment & Planning A* 43 (2011): 2001–2006.

Hawkins, G. Packaging water: Plastic bottles as market and public devices. *Economy and Society* 40 (4) (2011): 534–552.

Hawkins, G., and K. Race. Bottled water practices: Reconfiguring drinking in Bangkok households. In *Material Geographies of Household Sustainability*, ed. R. Lane and A. Gorman-Murray, 113–124. Surrey, UK: Ashgate, 2011.

Potter, E. Drinking to live: The work of ethically branded bottled water. In *Ethical Consumption: A Critical Introduction*, ed. T. Lewis and E. Potter, 116–130. London: Routledge, 2011.

Race, K. Frequent sipping: Bottled water, the will to health and the subject of hydration. *Body & Society* 18 (3–4) (2012): 72–98.

Gay Hawkins, Emily Potter, and Kane Race
Sydney, December 2014

Introduction: Markets, Materiality, and Biopolitics

In 2004, New York comedian Lewis Black stood on a Broadway stage and delivered a long rant about bottled water. It began like this:

We care so much about health that during the last twenty-five years we destroyed water. Because when I was a child this was the simplest thing of all, this was the essence of life. And when you were thirsty—and that's the operative word, ladies and gentlemen—the word is and always will be "thirsty" not "hydrate," they made that fucking word up! I could go anywhere in my house, I could go to three or four different rooms, I could even go to the basement and get clean water and drink it … mmmmmh … and then go back out and play, and those were great times. But then they decided that every town and village that had water coming to it and all they had to do was clean it: "We'll save money, we won't clean the water so much and with the money we save we can then buy the water at the supermarket." Try to go through this logic with me: our country had water coming to our homes and even if we were locked out we could still get clean water and we said: "No, fuck you! I don't want it to be so damned convenient, I want to drive and drive and look for water—just like my ancestors did." So now we buy water from Coke and Pepsi because when I think clean water, well yeah, I fucking think Coke and Pepsi…. Aquafina? I think that means the end of water as we know it. (Black 2004)

A transcript doesn't quite capture the affect of this performance, nor the emotional force of a lone man standing on a stage haranguing the bottle of water in his left hand. For at the heart of this rant was a barely controlled rage about what bottles and brands and beverage corporations were doing to water—how they were turning "the essence of life" into yet another market segment.

Global beverage companies are easy targets. Lewis Black was making a standard critical move, and audiences familiar with boycotts and consumer citizenship were predictably responsive. But it was the other registers of his rant that were more significant, and that often got the biggest laughs: the nuanced, almost ethnographic observations about how plastic bottles of water had insinuated themselves into everyday life. Black's account was historically aware, richly empirical, and infused with an implicit understanding of bottled water as much more than a demonized commodity. The descriptions of how taps dotted about the house configured water as readily accessible, the account of how the health information on labels medicalized the experience of thirst and turned it into "daily hydration requirements," the fact that no one seemed able to go anywhere without carrying a bottle and constantly sipping from it, the long drive to the supermarket searching for water—all these observations made their appearance in his presentation. By sticking closely to the bottle, Black documented some of the myriad new practices and relations in which bottled water was caught up, and the ways in which this innocuous product was interfering with existing forms of potable water provision. By the end of this routine he had very effectively explained the emergence of a new water reality.

When we first watched this performance on YouTube early in our research, there was a moment when our project seemed redundant. Lewis Black, it seemed, had said it all. Then we began to think *with* Lewis Black: we began to let the affective intensity and pleasure of laughter extend our critical sensibilities. Stand-up comedy, it became clear, was as important a form of case study as any other in understanding the recent emergence of mass markets for packaged water and the growing numbers of people carrying it as a personal accessory. Not only did Lewis Black's comedic routine reveal the ways in which new markets and consumption practices were vulnerable to potent ridicule, when the YouTube clip started to get thousands of hits and comments, it was obvious that this rant was contributing to the formation of a growing number of controversies and publics concerned with the impacts of turning water into a product. Comedy could not just satirize and unmask, it could also be a generative political event. Equally compelling was Black's technique of close observation. Driven by a fascination with what bottles were actually doing in the world and the new practices and beliefs they prompted, it was implicitly ontological. Reality wasn't just out there, it was enacted, and mundane things like plastic bottles were important participants. Once we acknowledged the methods and reverberations of this stand-up routine, Lewis Black's wasn't a threat but an example.

Where comedy finishes, this book begins. Like Lewis Black, we are interested in exactly how and why branded bottles of water have insinuated themselves into daily life and the implications of this for safe urban water supplies. What is at work in the movement from drinking from the tap to drinking from the bottle? This opening investigation led us to explore the development of new markets in water and the ways in which these markets have been constructed. What kinds of practical devices, calculations, and arrangements have been deployed to make water into an economic good? How has water been detached from other settings and requalified as a healthy convenience drink? We are also concerned with the novelty of increasing numbers of people sipping water from single-serve plastic bottles. When did it become necessary to carry a personal supply of water everywhere? How did water emerge as a form of portable property? How have these practices developed, and what diverse networks of attachment circulate between the product and the consumer? For a market arrangement is nothing without consumers who recognize the distinct qualifications surrounding the product and understand how to incorporate the product into their world.

Beyond questions of market organization and the dispositions and practices of consumers, another aim of the book is to investigate how bottled water has generated numerous "political situations" (Barry 2013). Lewis Black is an example of this—standing on a stage in New York, he made bottled water controversial—but there are many other controversies and situations in which bottled water has emerged as a political problem and a political object (Braun and Whatmore 2010)—witness the number of anti–bottled water activist campaigns that have stalked this product almost from the beginning of its rapid market expansion, or the growing plastic waste problems that discarded bottles are generating all over the world. We wanted to understand the processes whereby bottled water became controversial, and how its diverse framings have generated particular political conducts and effects. But this visible activism, this politics as organized contestation, is by no means the sum of the politics of bottled water. If bottles are implicated in new drinking and waste realities, how do these realities subtly (and not so subtly) interact with other realities? In what ways do the ontological dimensions of drinking bottled water—the ways in which this habit enacts new relations and forges new meanings—interfere with other drinking water practices? These questions underscore our concern with how bottled water is related to other systems for delivering potable water, and how it might animate wider debates about urban water provision and growing water scarcity. Obviously, bottles meddle with existing hydrological cultures and water infrastructures—or the absence of them— in vastly different ways in different places. Bottles are making differences that can be described as a form of ontological politics. Investigating these ontological politics is another central aim of *Plastic Water*.

The questions driving this book—about market assemblage, new practices of drinking, and diverse forms of politics—highlight the issue of multiplicity. Bottled water obviously has different meanings in different settings; its social life is complex and extensive. The same polyethylene terephthalate (PET) bottle of water can exist as a product, as a personal health resource, as an object of boycotts, as part of accumulating waste matter, and much more. How, then, should we attend to this multiplicity, recognizing the bottle of water as an object with an open and unfolding character and variable capacities for agency and efficacy in different settings? Although we use the generic term "bottled water," our approach is to situate the object, to pay close attention to the relations and associations in which it is caught up. These relations configure bottled water in particular ways, but the bottle is not passive in this process; it, in turn, affects those relations. Our aim is to pay close attention to these configurations, to attend to the dynamics of materiality, relationality, and process.

Awareness of multiplicity does not, however, deny the force or dominance of some configurations over others. Bottled water is most often described as a beverage product—or "fast-moving consumer good" (FMCG), to use the standard industry classification. What this classification captures is a particular and relatively new form of market organization for water that constructs it as a single-serve, packaged, immediate consumption product. As Tony Clarke (2007, 12) notes in the most definitive study of the industry to date, between 1993 and 2004 the consumption of bottled water as an FMCG doubled in the United States, to the point that it became the second largest commercial beverage sold. A 2010 Euromonitor report, *Drinking Cultures of the World*, provides further evidence of this significant market growth. It notes that the United States remains the biggest market in value terms, accounting for around one in every five dollars spent on bottled water globally. Other key findings were that bottled water was the most visible new force in the beverages sector and was challenging the dominance of carbonated drinks in a number of markets. In Europe, traditionally the hub of branded water cultures, sales had stalled in the wake of a "consumer rebellion against throwaway plastic bottles" (Euromonitor 2010, 14), but emerging markets in China, India, Brazil, and Indonesia were seen as the "silver lining for multinational water players" (16). Central to the dramatic market expansion in these emerging regions was the rapid development of packaging industries and supermarkets, both of which were seen as critical factors in market growth. Despite significant differences in drinking cultures and habits across the world and in the evolution of markets, bottled water was usually promoted as a product with a short shelf-life and high turnover, designed for consumption on the go. These specific details about the role of packaging, distinct consumer uses, and purchase settings are important. They highlight the diversity of elements and devices involved in making water into an FMCG.

Of course, water was marketized long before it became an FMCG. Beyond the long history of boutique mineral water markets, which dates back to the nineteenth century, there is also the significant role of small-scale water vendors and bulk suppliers selling water in twenty-liter containers in the many places where reticulated supplies are nonexistent or unsafe (Euromonitor 2010; Kjellén and McGranahan 2006). The "privatization" of public water services in many countries over the last thirty years is another development, as manifested in the various market transformations in place or under way (see Bakker 2003, 2010). These other water markets have distinct histories, geographies, and configurations, and although our focus is primarily on bottled water as an FMCG, its interactions with other market and nonmarket forms of water form an important part of our analysis. The key issue is that water can be turned into a market object in many different ways, so it is crucial to pay close attention to the specific processes whereby it is "rendered economic" (Muniesa, Millo, and Callon 2007, 3).

While bottled water's identity as an FMCG may be powerful, it is also unstable. Despite all the advertisements promising "natural purity" or "hydration on the go," these market framings and qualifications are subject to continued negotiation as the bottle circulates and is exchanged, consumed, refilled, thrown away, or becomes the object of any number of other actions. The point is not that we need to trace bottled water's changing identity and characterization through commodity chains, as if the structures and networks of capitalism always had the final say. Rather, the bottle of water should be approached, as we have already suggested, as an unfinished or entangled object (Thomas 1991), and the nature of these entanglements demands careful elaboration.

For Michel Callon (1998, 19), understanding entanglement is both an empirical and a theoretical project. It involves investigating how things get arranged or composed in particular ways, and how these arrangements come to have effects in the world or prompt certain actions without obvious connection to an underlying driving logic. "Assemblage" is another term that captures this dynamic process of arranging and ordering, and the contingency of these arrangements. In *Plastic Water*, we draw on the rich legacy of debates about the emergent and always changing conditions of assemblages to explore how bottled water's shifting social ontologies are constructed, become stabilized, and prompt various effects (see Bennett 2010; DeLanda 2006; Deleuze and Guattari 2003; Law 2004a). Although the organization of mass markets for bottled water is a central concern, we understand these markets as tentative and unfolding, and only partially under any form of deliberate control.

This focus on assemblage and its evocation of relations and heterogeneity may be unsettling for those who perceive the rise of bottled water in terms of the substantive power of corporations, capitalism, and neoliberal ideology. Not only do these elements represent major social forces, they also give shape to political opposition in the form of clearly identified structural causes for the marketization of water. The problem is that, in focusing on these macrostructures, complex processes of market emergence and *agencement* (Çalişkan and Callon 2010) often get reduced to a few determinants. Equally troubling is the way in which politics is represented as external to or outside markets. While there is no doubt that the dynamics of framing and the construction of calculative measures within markets generally seek to place politics outside the frame, in the realm of "externalities," this process is never complete. As Andrew Barry and Don Slater (2002, 185) point out, "far from limiting the possibility for political conflict and negotiation, framing forms something like a surface on which forms of political reflection, negotiation and conflict can condense."

It is not that we deny the presence and force of capitalism and beverage corporations; rather, we choose to look closely at the myriad elements, techniques, and concepts that are deployed to frame water as calculable, and the precise ways in which these framings become problematized and subject to multiple forms of contestation. Just because we resist the abstractions of big concepts and political analysis focused primarily on critique and moral condemnation, this does not mean that we abandon a political perspective on markets. Instead, we want to explore exactly how political situations have been practically assembled in relation to bottled water, and the ways in which they often emerge in processes of creative contamination with markets. If the rigid opposition between economics and politics is refused, then it is possible to see the conditions of their "clandestine hybridisation" (Cochoy, Giraudeau, and McFall 2010, 141), or the ways in which politics is immanent to many market forms.

Central to our approach, then, is a concern with how the political is performed and enacted in specific instances—how different actors deploy particular political categories and analyses in order to make bottled water a matter of concern. This way of understanding politics—as tactical, pragmatic, and situated—reveals a remarkable diversity of political actors and issues, an incredibly lively field of disputes and publics wrestling with the many troubling aspects of bottled water, from the container, to the contents, to the corporations that produce it. Little wonder this product has been described as "the new smoking" (Coren 2008)! It also makes it possible to investigate how political situations force markets to reflexively negotiate with innumerable actors, from ethical consumers to NGOs, as the product is redefined as a problem.

Our understanding of politics does not dismiss corporations or capital, but it does displace them from the center of the bottled water story. The performative and empirical approach to market assemblage that we develop here makes it impossible to reduce a multiplicity of participants and effects to the inexorable logic of a single element. It refuses to locate the cause of market development in a sole initiator or structure. Rather, it considers how *agencement* of a market—its capacity to generate economic and other actions (expected and unexpected)—emerges through the process of arranging various sociotechnical, corporeal, discursive, and material elements. Beverage corporations and flows of capital are crucial elements, but so too are many other mundane devices, formulas, and practices that help actualize and expand market relations. This is not to assume equality among all actants or to envisage a perfectly level field, but rather to insist that the agency of an assemblage is always distributed: alongside elements with powerful intentions there exists a "heterogeneous series of actants with partial, overlapping and conflicting degrees of power and effectivity" (Bennett 2010, 33).

In seeking to shift from a general political economy to a political perspective on bottled water markets, we begin this book by looking closely at three elements: the bottle, the water, and its suggested use, constant sipping. These choices are not random or indiscriminate. Our claim is that each of these elements and the complex relations between them affect economic action in very significant ways. The PET bottle was an innovative form of plastic packaging that was first used in the beverage industry in the early 1980s, with phenomenal impact. The explicit function of this market device was to reduce costs through reducing weight while retaining the durability and chemical stability of glass. What was latent or unexpected in this dramatic rematerialization of beverage packaging was PET's remarkable affinity with water: the way in which its material semiotics seemed to enhance water's biophysical translucence and its special claims to purity, and the way in which the fluidity and clarity of water were amplified by plastic's plasticity. But this powerful alliance between the packaging and the product was nothing without other transformative actions, such as the wider intensification of branding and shifts in consumer concerns about personal well-being. The rise of biocitizenship made it relatively easy to create a subject who could become attached to bottled water in the interest of a will to health.

This account is starting to sound very teleological: first came the bottle, then the water, then the anxious drinker, then the rapid expansion of a new market segment in the beverage industry. That is not our intention; instead, we consider these elements as crucial developments in themselves, but also in terms of how they interact—how they generate overlapping degrees of power and agency in market arrangements. While there is no denying the pressure for more and more things to be marketized, our concern is with exactly how this happened in relation to water—how these elements were implicated in the emergence of a new reality for water as an FMCG. Equally important is the way in which attention to these elements is able to override the dualism between subjects and objects, or between production and consumption. Rather than conceive of the consumer of bottled water as manipulated by advertising or possessed of false desires or anxieties about tap water, we consider how these three elements—packaging, branding, and new health discourses—are differentially and collectively implicated in qualifying water and generating diverse forms of calculation for it. Using Callon's (1998) definition of a market as a "co-ordinating device," or endlessly variable assemblage for the calculation of value and exchange of goods, we pay close attention to how bottled water markets organize relations between humans and nonhumans (materials, discourses, sociotechnical devices) and render everything involved in a market arrangement a calculating agent to different degrees. These calculations are realized not just in sales figures but also in the myriad ways in which bottles of water generate specific consumer attachments—how the drinker becomes a calculating agent in the actions of choosing and carrying packaged water. As Liz McFall (2009, 275) puts it, "the virtue of a sociology of market attachment lies not in telling us about consumption generally ... but in telling us about the distributed and material character of market processes specifically."

We investigate these three elements in part I of the book, "The Event of Bottled Water." Our approach to the event is both literal and philosophical. The rapid growth of mass markets in bottled water was not simply something that happened in progressive time, a novel development in the beverage sector, but it was also the actualization of new and often surprising relations between and among plastic bottles, water, and the drinking body. The importance of this approach is that it casts events as processes whereby things become different without recourse to reductionism. In seeking to understand how mass markets in bottled water were created, we move across time and space: from the invention of the PET bottle in 1973 in a DuPont laboratory, to practices of economizing spring water in nineteenth-century France, to the development of "hydration science" with the rise of jogging in the United States in the 1970s. All these elements participated in crystallizing the event of mass markets in bottled water: it would not have happened without them, but they weren't enrolled simply according to the neat logic of cause and effect. In this way, then, event thinking understands processes of market assemblage as both historical and contingent. Rather than explain rapid market development according to the logic of inexorable market expansion, we look at patterns of emergent causation and the ways in which developments in plastic packaging, the rise of branded waters, and new health discourses interacted with and rebounded off each other. And while these elements enabled literal market growth, it is their contingent coming together and the proliferation of effects that are significant, and that signal the philosophical significance of the event. Event thinking also points up the distributed nature of agency: the way in which new relations between things make a difference. While there is no doubt that bottled water has made innumerable differences in everything from how we understand water quality to the expanding amounts of plastic waste, these differences, these real and serious effects, must be understood as both causal and emergent, as predictably shaped by powerful forces such as beverage companies' interests but also accidental and unpredictable and extensive.

In part II, "Bottle Practices," we consider the extensiveness of the event of bottled water market growth by investigating what bottles do in the world. Our primary aim in this section is to understand how bottles of water have been actualized as *ordinary:* how bottles function practically, how they acquire meaning, and how they become normalized as a routine feature of everyday drinking and disposal practices. Our focus is on three Asian cities, Bangkok, Chennai, and Hanoi. These sites allow us to investigate bottle practices in regions where markets are growing fast and where access to reliable reticulated water is nonexistent or unsafe. In these cities, splintering urbanism is the norm, and our concern is to understand how bottles become participants in fragmented water supply and waste infrastructure networks. These messy and highly situated hydrological and urban realities have significant implications for how bottled water markets are assembled and the ways in which water in bottles is both qualified and consumed. Not only does bottled water often become *re*qualified from an FMCG to an essential component in daily survival and household practices, it also helps define new water hierarchies, in which different sources of water are used for different purposes.

The detailed empirical accounts of bottle practices that we develop in part II problematize notions such as "privatization" and "neoliberalism." While there is no doubt that the rapid rise of markets in the cities we examine is an opportunistic response to various forms of state failure and fragile water environments, oppositions between public and private do not come close to capturing this complexity. It is only by seeing how bottled water is made meaningful in relation to other forms of water provision and everyday practices that it is possible to understand both its ontological realities and its ontological politics. To this end, we investigate how bottled water is being incorporated into middle-class households in Bangkok in response to concerns about the municipal supply, how it is implicated in both constructing and addressing water scarcity in Chennai, and how accumulating discarded bottles are generating new practices of economization in the form of plastic recycling villages on the outskirts of Hanoi. We are concerned with the dynamics of social practices, with the ways in which bottled water is implicated in assembling and enacting new drinking and waste realities. The key issue we wrestle with in this section is how bottles generate what we call "shadow" political effects, whereby market forms of delivering potable water become normalized as safer or more convenient or more aspirational and implicitly undermine the struggle for the development of public or nonmarket forms of safe drinking water provision.

Finally, in part III, "Ethical Drinking," we investigate how bottled water has become politicized or made into a "hot situation," to use Callon's (2008) well-worn phrase. Here, political effects are by no means shadowy, nor are they operating simply in the registers of ontological interference. Instead, the impacts and externalities of bottled water markets are made explicit through the use of counterinformation, NGO activism, and minoritarian issue networks, among other mechanisms. Our interest is in the diverse political situations that have emerged to contest the uses and effects of bottled water and the ways in which these situations have evolved. We take our bearings from recent debates about material politics (Barry 2013; Braun and Whatmore 2010; Marres 2012) that focus not only on the role of numerous materials and devices in constituting political action but also on the increasing proliferation of controversies about the material things and objects that call concerned publics into being. Through an investigation of three different political situations—the Inside the Bottle campaign run by the Canadian Polaris Institute, the "FilterForGood" marketing strategy developed by the Brita water filter company in the United States, and the reintroduction of public water fountains in an Australian city—we explore a couple of key questions. First, how did these situations frame bottled water as a political problem—what aspects of it were rendered ethically troubling? And second, what practical devices and techniques were deployed to assemble publics willing to say no to the bottle?

Each of these campaigns implicates a wide range of people as "communities of the affected" (Marres 2012). They show how bottled water is not simply the problematized object of political deliberations and activist critique; it is also a political actor—something that can acquire lively "powers of engagement." This is Noortje Marres's term, and she uses it to describe forms of contemporary political participation that have moved beyond the dynamics of simply informing citizens about troubling issues to the active implementation of changes in the material practices in daily life. In these situations, engagement is both embodied and more-than-human. Engagement requires specific objects, technologies, and practices in order to invigorate political participation and make publics. The challenge, then, is to empirically track the actual processes of assembling bottled water issue networks, to document how the bottle becomes something with the lively capacity to engage consumers as publics and ethical drinkers.

But what of the effects of these organized forms of contestation on bottled water markets? How do they respond to these overflows and manage the significant challenges they pose to market calculations? In chapter 8, we explore these questions by looking at ethically branded bottled water and the rise of cause-related marketing by leading beverage companies. Although it is easy to dismiss such developments as little more than greenwashing or the relentless incorporation of externalities, we resist this assessment. Instead, we consider how the rise of what Barry (2004) terms "ethical capitalism" highlights the reflexivity of markets and the ways in which corporations seek to act on their business activities and display ethical conducts. This is akin to Callon's (2009) argument that some markets become "civilized" by responding to various issues and problematizations through specific sociotechnical developments. Callon's claim is not that markets can solve everything or incorporate all opposition but rather that they can use contestation and political situations as resources for experimentation in new market forms.

Our aim is to understand what business conducts and activities have been deployed to make bottled water "ethical." How does a corporation make visible or demonstrate its ethical concerns? And in what ways do these practices shape new relations between products and consumers that generate various forms of ethical value or capacity in the bottle of water, the drinker, the brand, the beverage company? However, the processes by which a market develops specific calculative capacities to enact ethical forms of action and value are only part of the story. We are also concerned to understand how the event of bottled water activism, the emergence of significant critique and opposition, is implicated in these developments.

In what sense do forms of market contestation and controversy cease to be externalities and become instead unavoidable elements in the architecture and calculative equipment of a market?

At the heart of *Plastic Water* is a desire to understand how mass markets in packaged water and new practices of drinking have been created, and the multivalent political effects of these developments. Although we are skeptical about explanations that reduce these complex processes to expressions of large-scale forces or grand problematizations, this does not mean abandoning a critical position and its potential for inventive scholarly intervention (Braun and Whatmore 2010, xxvii). As we have outlined, a central concern is to develop critical explanations of these market and everyday assemblages, to understand how they generate a multiplicity of political situations, interferences, and disturbing shadow realities. But what is the basis of our critical position? How do we judge these political effects as not simply tactical or performative but also implicated in more universal ethical concerns about what bottles are doing to water and its fundamental role in sustaining life?

The unavoidable context for the rise of bottled water is biopolitics, a concept Foucault deployed to describe the myriad ways in which biological life becomes an object of political calculations (Agamben 1998; Foucault 2008). The basis of our evaluations of the effects of packaged water begins in the centrality of water in sustaining life. As Karen Bakker (2003), Matthew Gandy (2006), and many others have argued, water is a "nonsubstitutable resource"; it is fundamentally implicated in diverse forms of rule, population management, and the various lines of connection and interdependence between human existence and nonhuman nature. Access to potable water for growing world populations and the protection of water sources and fragile hydrological environments are among the most pressing issues of the present. These global realities form an essential background to our investigations. While we do not review them directly, our aim is to understand how markets in packaged water often animate and amplify them by making present—implicitly or explicitly—wider questions about the value of water and its place in enabling life. In this way, we seek to understand how very specific and historically contingent forms of biopolitics are enacted not only in the emergence of new market forms in water but also in the frequent marketing claims that accessing the bottle equals accessing life.

Situating the phenomenal recent rise of bottled water in relation to this biopolitical context means taking the active material presence of bottles seriously. The proliferation of plastic bottles of water virtually everywhere in the world cannot be read as an indicator of generic biopolitical or epochal shifts, such as the rise of neoliberalism or risk cultures. As we will show, this veritable swarming of plastic bottles is the outcome of highly situated provisional alliances and makeshift arrangements in which certain conducts and materials (the products of the plastics industry or water from natural springs, for example) come to make convenient sense, and generate what Stephen Collier and Andrew Lakoff (2005) describe as new "regimes of living." This term explains the emergence of situations in which "living" is rendered problematic by subjecting mundane acts such as drinking to shifting norms, reasoning, and material practices. It also indicates the increasing centrality of biological processes in new practices of economization, and the ways in which changing bodily conducts are materialized through complex networks of the human and nonhuman. Our aim, then, is to track how plastic bottles emerge as participants in distinct regimes of living—how they are enrolled in new economic and technical arrangements for delivering water, and how they can become capable of undermining trust in public water provision or derailing the ongoing struggle for it. This approach makes it possible to examine bottled water without putting humans or corporations at the center of the story. It begins from the modest recognition of plastic bottles not as phobic objects but as things with which we are caught up, things that are materialized through diverse economic processes and shifting everyday habits. It also enables an understanding of how the bottle performs in different arrangements to both contain and frame water, how it becomes implicated in changing the qualities of water and inviting new forms of ethical reflection about drinking and ways to live, and how it articulates new relations between capital and life.

I The Event of Bottled Water

1 Packaging Water

Unlike many commodities, bottled water draws attention to its packaging. The ubiquity of water as a reticulated service, as something that flows (at least in the global north), is disrupted by foregrounding the mode of delivery. The bottling of water turns an ordinary liquid into a mobile commercial beverage. The bottle doesn't simply contain the water; it makes it available for new forms of branded exchange and new practices of drinking. Acknowledging that the bottle matters in the emergence of mass markets in water may seem obvious, even banal, but the nature of this mattering is at the heart of the bottled water phenomenon; hence our decision to start with the bottle. For this small, mundane plastic container is both pragmatic and potent. It has not simply allowed water to be distributed and consumed in different ways, it has also made it possible for new forms of economic life to emerge around this essential liquid. How, then, to understand what bottles do in the world?

For many commentators, bottles cause nothing but trouble. While the activist literature on the rapid rise of bottled water markets recognizes the centrality of the bottle, the bottle's role is often reduced to that of a passive instrument in the corporate commodification of water. In these critiques, the issue is not so much the bottle but what it contains. Although plastic bottles have long been a feature of beverage markets, filling them with water has prompted a massive range of activism.[1] Since the development of state water infrastructures in the nineteenth century, access to safe water has been emblematic of citizenship, and the rise of mass markets in packaged water has been criticized for undermining this relationship between state and citizen, for threatening the biopolitical framing of water as a state responsibility and public good.

Beyond this fundamental concern about turning water into a market good, other critiques focus specifically on the effects of the packaging. A range of technoscientific and environmental disputes has emerged around the increased use of plastic bottles and the impacts of such use. The leaching of chemicals into the water, the phenomenal growth in plastic waste around the world, and the unsustainable practice of moving plastic bottles vast distances from water sources are a few of the many problems that have been highlighted. In making bottles into a matter of concern, activism has turned this common container into a demonized object.

But exactly how have bottles reconfigured the qualities of water, its distribution, and everyday drinking habits? The great value of critical commentary is its documentation of the multiple and proliferating effects of bottles in the world. Critics and activists have implicitly captured the sense of bottles as "unfinished objects"[2] having various futures as packaging, container technology, litter, objects of protest, or contaminants, among other possibilities. However, the limitation of critical commentary is its tendency to reduce such multiplicity to one or a few determinants. In the influential study *Inside the Bottle* (Clarke 2007), for example, the big four global beverage companies, PepsiCo, Coca-Cola, Nestlé, and Danone, are placed at the center of the bottled water story. They are presented as the most powerful agents in fashioning and expanding bottled water markets. These corporations are seen to be acting in a separate sphere—"the economy"—and the markets they create are assessed as amoral and unilaterally exploitative of both natural resources and consumers. The impacts of the bottle on water sources, waste management systems, and drinkers are inevitably traced back to corporate intentionality. Despite its multiple ontologies, the bottle becomes an object with an unfolding logic already embedded within it: it is an instrument for capital accumulation. As much as critical discourses are concerned about bottles, they are strangely blind to them as concrete material things. In critical narratives, bottles have innumerable effects, but they are incapable of articulating action. They are rendered problematic but inert, excluded from the active realm of the social (Barry 2005).

Rather than reiterate this critical commentary and the assessment that bottles are essentially bad, our aim is to investigate them as multiple and active objects. This doesn't mean we aren't troubled by their effects; rather, we want to understand how these effects develop, how they emerge in the specific associations in which bottles become entangled. Our concern is to develop a less reductionist and more *plastic*, so to speak, account of the emergence of bottles as packaging for water, as personal accessories, as environmental problems, and so on. According to Manuel DeLanda (1995), the actual material properties of things do not ensure any kind of essential or singular function, for these properties have a "multiplicatory role": they have various potentialities or capacities according to the shifting associations within which things become caught and the ways in which they perform in these associations. Attention to bottles' multiple roles, then, does not deny the stubborn persistence and dominance of some bottle realities over others, particularly the bottle's ubiquity as beverage packaging. However, it does highlight the ways in which these dominant realities are ontologically unstable and the ways in which the actual entity of the bottle in one setting is always capable of giving way to other realities. Bottles, then, have to be understood as actual *and* processual, and the task for analysis is to explain the processes whereby the various functions and values of bottles emerge, are stabilized, and can also change.[3]

In this chapter, we take seriously the "bottle" part of bottled water, and investigate not only the ways in which plastic bottles have become the dominant packaging in the beverages industry but also how they have formed particularly powerful alliances with water. Polyethylene terephthalate (PET) bottles contain innumerable commercial beverages, but their relationship to water is distinctive and central to making water calculable. Unlike sweetened beverages and juices, water is already easily and cheaply available in everyday life for many people. Making water into a market entity requires significant effort to separate water from its ordinary context and establish special or unique qualities for it. In this way, the bottle does far more than contain a common fluid: it is also implicated in conveying new meanings for water and in organizing commercial actions; it is both technical and social. The derisive claim that bottling water marks the ultimate triumph of packaging acknowledges this process and its cultural effects. It alludes to the ways in which the container is far more than a functional device; rather, it is something that gathers many new elements and relations together that fundamentally transform the contents.

Underpinning this initial focus on the bottle is a recognition that the phenomenal growth of markets in packaged water over the last twenty-five years must also be considered an *event*, a development that has signaled a proliferation of new relations and networks for water (Harman 2009, 64).

The packaging of water by beverage companies generated novel effects and repercussions: bottles became launched on multiple trajectories, many of which interfered with existing practices of water provision and drinking. Bottles formed alliances with a whole range of things, from aquifers to the thirsty body to branded refrigerators in supermarkets. They became participants in various assemblages and influenced how these were arranged; they acted in certain ways and prompted other creative and unpredictable emergences, from informal plastic recycling efforts to online activism. So, if the relatively recent growth of bottled water markets and consumption is an event, making sense of this event means acknowledging bottles as critical actants, as sources of action, as things that make a difference (Bennett 2010, 10).

Tracing how plastic bottles became actants, in beverage markets and beyond, involves careful attention to the ways in which their specific material capacities emerged and acquired distinct forms of agency in different settings. Our primary concern is to investigate how bottles acquired *economic* capacities, how they participated in the formation of new economic processes around water and became crucial market devices. What was the role of plastic packaging in assembling expanded beverage markets in water? In what ways did the particular material-semiotic qualities of plastic become implicated in the emergence of new qualities for water and new drinking practices? Of course, many other objects, institutions, regulatory processes, and technologies were involved in processes of market organization. Bottles didn't work alone—but in foregrounding this material object in this chapter, our aim is to shift thinking away from the idea of packaging as instrumental to packaging as performative, as something that helps bring new realities and practices into being that have socially binding effects (Butler 2010).

As we have indicated, our claim is that packaging water in single-use PET bottles can be understood as an event that prompted complex trajectories and effects. These effects were predictable and unpredictable. They reverberated from the bottle outward and continue to reveal its extraordinary heterogeneity and complexity. It is not that economic or technical or cultural processes didn't influence this event—they played a crucial role in patterning it, in shaping the form the bottle took and how it became enrolled in various assemblages. But the emergence of the plastic bottle cannot be reduced to these processes. The bottle isn't simply an instrument of economic forces with inexorable logics, nor is it a mere expression of technical advances in plastics research. It emerged out of the interactions of industrial, commercial, and creative processes that were highly situated. And central to these processes were the properties and "expressivity"[4] of plastic, the ways in which it made its presence felt. In tracing how the bottle became a market device in the beverage industry, then, we begin with an investigation of the behavior and vitality of PET, the form of plastic that is most commonly used in bottle packaging. Our aim is to trace how PET's synthetic potentialities emerged and how the PET bottle became an "economically informed material," that is, something capable of expanding and intensifying market logics in meaningful ways. Sticking to the bottle, understanding its processes of emergence and the contingency of it effects, is not to replace economic reductionism with material reductionism. It is simply to insist that, in the event of bottled water, the packaging was a powerful actor in the construction of new economic actions around water and new ways to drink.

The Plasticity of Plastic: Inventing the PET Bottle

There is little question that the growth of PET packaging has been the big success story in packaging in the 1990s. It now appears that this story will continue to dominate our attention well into the future. PET has demonstrated an unusually strong connection with consumers. For whatever reasons—portability, lightweight, convenience, safety—PET packaging is helping brand owners sell product.
—Giles and Bockner 2002, 26

Beyond the celebratory, almost promotional tone of this account, Geoff Giles and Gordon Bockner make an important point: developments in PET reveal the impacts of massive transformations in the evolution of packaging as essential infrastructure in a multitude of markets. Not only has PET replaced other materials; it has also prompted new forms of packaging and the expansion of applications "well into the future." While Giles and Bockner in the passage cited focus on the distinct appeal of the properties of this material to consumers, those same properties—light, portable, safe—meant that it was also immensely appealing for packaging applications in food and drinks. In beverage markets, the shift to the PET bottle was particularly rapid. While U.S. commercial production of PET bottles was negligible in 1977, by 1980 some 2.5 billion bottles were being produced, and by 1985 production numbers were up to 5.5 billion units per year (Plastics Academy n.d.). The development of the PET bottle made a significant difference to the organization and development of beverage markets and heralded shifts in both production and consumption. This was obviously a plastic bottle like no other before it, and its invention and material performance warrant close attention to understand how the PET bottle became such a potent element in the organization of beverage markets.

The standard story about the development of the PET bottle goes something like this: by the early 1970s, blow-molded thermoplastic bottles had successfully replaced glass in most household containers for everything from shampoo to detergents. However, application to the beverage industry proved difficult. The thermoplastics used for bottling detergents and other nondrinkable fluids were considered unsuitable for carbonated drinks and fruit juice because the contents tended to attack the plastic. This material instability led to a range of problems, from deterioration, to explosion of the bottle under pressure of carbonation, to chemical contamination of the contents (Freinkel 2011, 172). The beverage industry therefore presented a vast field of possibility for the expansion of plastic bottle packaging. This was an industry dominated by glass bottles and, since the early 1960s, aluminum cans. Although the development of the single-use aluminum can had had a major effect on the growth of "on-the-go" markets—particularly Coca-Cola's and Pepsi's decision to diversify into them in the late 1960s— there was still considerable interest in finding a stable plastic that could be used in the beverage market. Throughout the early 1970s, Coca-Cola and Pepsi tried bottles made of acrylonitrile resin, but in 1977 the U.S. Food and Drug Administration (FDA) revoked its approval when detectable residues of the monomer were found to have migrated into the contents. According to David Brooks and Geoff Giles (2002, 2), this regulatory move by the FDA was a powerful impetus for the "PET bottle revolution," because it prompted further investigation of other plastics.

Regulation was effective in eliminating some plastics, but the issue of how to find the best plastic to safely containerize beverages remained. Around the same time that acrylonitrile was in test trials, DuPont was working with Pepsi on the bottle possibilities of PET. Since the 1960s PET, a thermoformed polyester, had been used to make molded food trays and packages, but turning it into bottles proved elusive. PET could be used to make the preform base of a bottle, providing essential stability, but not a whole, narrow-lip, hollow container. Engineers at DuPont had long been working on developing a suitable plastic and production process that could mold PET into a full beverage bottle. An object pathway was in place, but a material innovation was needed to pursue it. The innovation that was developed in the DuPont laboratory was a new process of biaxial—or stretching in two right-angled directions—molding. While plastic bottle production had always involved stretching or blowing processes, the DuPont laboratory experimented with this technology using polyester resin. It was the first research facility to blow and stretch a PET bottle from an injection molded PET preform (Brandau 2012). These experiments eventually produced a "biaxially oriented polyethylene terephthalate bottle," as the 1973 patent defined it. This production process conferred remarkable new properties on the plastic. It stretched or reoriented the polymer chains of polyester, changing their macromolecular properties to produce a plastic bottle of incredible strength, lightness, and transparency. This bottle was very suitable for beverages. Light and virtually unbreakable, it had reduced permeability to carbon dioxide, oxygen, and water, as well as optical qualities that rivaled those of glass. As Susan Freinkel (2011, 172) explains:

Here was a plastic bottle that was tough enough to withstand all that pressurized fizz but also safe enough to win approval from the FDA. It was as clear as glass but shatterproof and just a fraction of its weight. Its thin walls kept out oxygen that could spoil food contents while holding in that expensive carbon dioxide. The PET bottle was yet another of those pedestrian plastic products that humbly fulfilled a Herculean set of demands.

In the trade press, the development of the PET bottle was attributed to Daniel Wyeth, a DuPont engineer, who dined out for years on the moment when, after months of frustration and failed attempts with molding polyester resin, he opened the mold, thinking it was empty, only to discover, on closer inspection, a crystal-clear plastic bottle (Brooks and Giles 2002, 1). Beyond the celebration of Wyeth as the individual "discoverer" of the PET bottle, the trade press documented the research-and-development (R&D) context of industrial chemistry, the multitude of factors and processes that had gone into the invention of this bottle over time, from technology transfers, to failed trials with other plastics, to diverse government regulations controlling everything from packaging safety standards to patents.

A focus on invention rather than on sudden discovery underscores the crucial role of R&D contexts and the ways in which they are shaped by what

Andrew Barry (2005) terms "operational realism," or highly situated instrumental and empirical logics. Finding a suitable plastic for beverage packaging applications was the goal of many industrial laboratories throughout the 1960s. The research effort was driven on the one hand by the technical limitations of existing forms of packaging, and on the other by the desire to capture a lucrative and growing market. The ideal plastic beverage bottle was a singular case, an object that guided research and posed questions to existing plastics and existing beverages containers.

In this way, the research effort was shaped by both external and internal considerations. It was concerned with the economic and social limits of existing beverages packaging and plastic bottles in use *and* with the internal structure of PET: how it behaved and what its material possibilities might be. These different scales, from the industrial to the molecular, from the production technologies of biaxial molding to how particular polymer chains might perform in response to new forces, suggest how the properties of materials are mediated by specific institutional settings and relations with other material and nonmaterial entities (Barry 2013, 13). They also accord with Bernadette Bensaude-Vincent's philosophical account of plastic. According to her, the defining feature of plastic as a material is its plasticity, and specifically the way in which the process of polymerization, which brings together the raw materials and heats them, occurs simultaneously with molding the material into a distinct object: "In more philosophical terms, matter and form are generated in one single gesture" (Bensaude-Vincent 2013, 20). Perhaps more than any other discipline, chemistry has developed an understanding of materials in an industrial context in response to the demands for products that could fulfill certain functions. The plasticity of plastic meant that material innovation, or "made-to-order" materials, often followed the definition of the ideal product.

This industrial context for chemistry produces what Bensaude-Vincent and Isabelle Stengers (1996) have called "informed materials." They describe an informed material this way:

Whether functional or structural, new materials are no longer intended to replace traditional materials. They are made to solve specific problems and for this reason embody a different notion of matter. Instead of imposing a shape on the mass of material, one develops an "informed material" in the sense that the structure becomes richer and richer in information. Accomplishing this requires detailed comprehension of the microscopic structure of materials, because it is playing with these molecular, atomic or even subatomic structures that one can invent materials adapted to industrial demands. (Bensaude-Vincent and Stengers 1996, 206)

Barry (2005) extends this concept of informed materials very productively. In his argument, industrial chemistry R&D does more than mechanically reshape or develop new materials for application in various fields. It invents informed materials in which the molecular is always constituted in complex informational and social environments. This approach to invention understands it not as progressive evolution or as the discovery of a fixed inherent potential but rather as a process of making materials *more and more informed*. In this process, the environment is not external to the material but enters into its constitution. As Barry (2005, 59) says, "The perception of an entity (such as a molecule) is part of its informational material environment."

Developments in post–World War II thermoplastics had established bottles as one of the key objects whereby information about plastics' material performance and economic possibilities was developed. Molding technologies, colorized plastic, and the squeezable bottle all played a role in revealing the diverse capacities of thermoplastics in bottle form.[5] The objects and materials informed each other (Shove, Trentmann, and Wilk 2009, 101). The invention of the PET bottle generated new material information about the performance of plastic and the possibilities of bottles. It appeared to enact the "bottleability" of plastic better than any other before it.[6] However, while this invention was patterned by technical and economic structures already in place in industrial plastics research and production, it also actualized surprising new properties for PET in the form of a chemically stable, lightweight bottle. In this way, the emergence of the PET bottle was an event, a novel occasion. The material capacities that emerged when polyester resin was stretch blow-molded into bottle form revealed the rich informational environment that surrounded plastic packaging research at the time, *as well as* the virtual potential of polyester, or what Manuel DeLanda (1995) calls the "expressivity of the material." In event thinking, the singularity of the PET bottle does not undermine the continuities of context, history, and existing knowledge, but neither is it reducible to them. As Gilles Deleuze (1993) has argued, the relation of events to existing states of affairs points up the dynamics of the actual and virtual. Events recognize the power of the virtual and the capacity of things to become different beyond the logic of continuity.

Market Devices: Making the PET Bottle Economically Informed

If the emergence of the PET bottle revealed new properties and the capacity of polyester to behave differently, what was the scope of this event? How did it make a difference to beverage markets, drinking practices, and new valuations of water? Industrial R&D and the expressivity of PET actualized a new bottle, but it was the way in which this bottle was enrolled in beverage markets as a new form of packaging that was more important. For it was in the transformation of the PET bottle into a commercial object that its capacities as a market device were not so much realized as enacted. How, then, did the bottle become a commercial object? How did it interact with existing market arrangements and rearrange them in particular ways? And how did it prompt new attachments in consumers?

In this section, we investigate the dynamics of the PET bottle as a market device within the beverage industry. Our focus is not exclusively on water markets, for the PET bottle was dominant in the marketplace before sales in packaged water started their rapid expansion in the mid-1990s. In the following section we consider the specificities of the PET bottle as a container that transformed water and established a very distinctive relationship with this ubiquitous liquid. Our particular concern is with how this plastic packaging played a key role in singularizing and qualifying water, and how these processes are coelaborated by consumers. In reaching for the PET bottle of water, the consumer apprehends a bottle as translucent as the liquid, a bottle that appears to simultaneously flow and contain, that seals in purity with the unbroken protection of a cap. These consumer assessments are not the result of manipulation and the conjuring up of false or misguided desires; rather, they are part of the way in which the PET bottle generates new properties for water and new networks of attachment that either resonate with the reality of the drinker or suggest better ones. In seeking to understand the complexities of this plastic container as a market device, the driving question we are concerned with is how the evolving intricacies of agency for this object have emerged and played a constitutive role in transforming how water is economized. To put it bluntly, could a market in bottled water have emerged without the event of PET?

According to Fabian Muniesa, Yuval Millo, and Michel Callon (2007), a market device is something that articulates economic action, always in relation to other devices. Market devices, then, are objects that *do* things.

Whether in a minimal instrumental fashion or a forceful determinist fashion, they are things that act and make others act. This capacity for action is not something possessed by the device, it is an outcome of the distributed agency that emerges in market assemblages. Koray Çalişkan and Michel Callon (2010, 8–9) extend this analysis with an account of *agencement,* a French term that describes a combination of heterogeneous elements that have been adjusted to one another. Embedded in the concept of *agencement* is the notion that agencies and arrangements are not separate, which suggests it is possible to trace the variety of forms of action all the entities in a market can generate. More important, the concept of *agencement* also makes it possible to understand how relations of domination are dynamically established.

This focus on devices and *agencement* shifts attention away from reductionist notions of cause and effect, or linear notions of straightforward market application, and toward the dynamics of what Jane Bennett (2010, 33) calls "emergent causation." Emergent causation refers to the actual processes whereby agency emerges and circulates. Rather than see the take-up of the PET bottle in beverage markets as simply causing market expansion, this approach maps the multiple effects and spatiotemporalities of the bottle as an actant with emergent capacities rather than fixed or predetermined impacts.[7] Tracking the actions of the PET bottle in beverage markets means paying close attention to the ways in which this plastic became enmeshed in new relations and generated effects that were both expected and surprising. It also means investigating how these effects became implicated in further changes by means of feedback loops. Emergent causation understands the bottle as an object with contingent agentic capacities that are constituted in particular market *agencements,* as well as in diverse uses, but that also *shapes* those relations with shifting degrees of power and effectivity (Bennett 2010, 33).

For the PET bottle that emerged from the DuPont laboratory in the early 1970s to become a successful packaging product, it had to go through various "qualification trials" (Callon, Méadel, and Rabehariosa 2002, 198). Callon and colleagues describe qualification trials as a sequence of transformations involving different networks and agents that the product singles out and binds together. The effect of these trials, through processes of adjustment and iteration, is to establish the characteristics of the product. In the case of PET, the function of the trials was to stabilize its material properties and render them predictable; only then could the bottle become mass-produced packaging and an effective market device. These trials involved continued interactions between the plastics industry and both packaging design and bottling companies. Throughout the late 1970s, aspects of PET that resisted full commercialization were researched and altered. The earliest PET bottles were slightly permeable to atmospheric gases, so small amounts of oxygen could enter the bottle. This was enough to spoil the taste of fruit juices and other beverages and limit their shelf life. In addition, the bottles could not be hot-filled because higher temperatures caused some breakdown of the plastic (Freinkel 2011, 172). All these issues had to be resolved before the economic capacities of the PET bottle could be realized. One of the most critical elements in PET's articulation in markets was standardizing it: controlling its material expressivity. To make PET commercially viable, it had to become a well-disciplined material. This didn't mean that PET was rendered permanently docile, for it was still capable of exceeding market regimes and prompting new patterns of *agencement* and emergent causation. Rather, stabilizing and standardizing the properties of the plastic were necessary to making it *calculable*.

Once the material properties of the PET bottle were stabilized, it was rapidly scaled up and became a mass object. The existence of a well-established plastics industry was central to this process of escalating mass production. Thanks to the "plastics explosion" during the 1950s and 1960s in the United States, there were plenty of production facilities, making the raw materials and feedstock necessary for numerous types of plastic manufacture. During the 1950s, the source of chemical feedstocks for the industry had shifted from coking coal to by-products of the petroleum and natural gas industries. As these industries expanded with the growth of the oil economy, so too did plastics production (Meikle 1995, 177). Jeffrey Meikle's account shows how the production of synthetic chemicals rose phenomenally as the petrochemical industry became more and more implicated in the economic logics of the continuous flow oil refinery. Petrochemicals were investment-intensive industries driven by a focus on constant high-volume production. As Meikle argues, "successful competition with labor-intensive industries like wood, wool or leather required pushing volume to the limit and finding a use for every by-product" (265). Disposable, single-use plastics used predominantly in food packaging were one of the key places where this volume was absorbed. The increased use of plastic packaging after World

War II was critical in helping the plastics resin industry expand and grow. But it was the development of the PET bottle in the late 1970s that signaled a dramatic increase in the production of plastics resins for packaging (266).

Another key factor in the rapid growth in PET bottle production was its relative cheapness and ease of production. Unlike aluminum can and glass bottle production, many beverage companies were able to incorporate PET bottle production into their filling plants, thereby cutting out interactions with packaging manufacturers. PET resin was readily available, as was the stretch blow-molding technology. Eventually, integrated systems to blow and fill the bottles at the same time emerged so that the beverage producer could control every element of the process, creating the container and inserting the contents in a single assembly line (see figure 1.1).[8] According to Tony Clarke (2007), by 2004 the three largest users of PET resin in the United States were Coca-Cola, Pepsi, and Nestlé, which among them were purchasing 80 percent of production.

Beyond ease of production and scalability, also crucial were the material qualities of PET, and the way its plasticity could be played with. For a start, the PET bottle was exceptionally light, strong, and flexible. This meant that

Figure 1.1
Mass production of bottles.
Source: Alamy/moodboard.

the dynamics of light weighting could be deployed to reduce the amount of material used, as well as the costs of transportation. In terms of design, the PET bottle could also be freely shaped into an enormous range of styles to enhance shelf appeal while still retaining functionality, and its surfaces could be directly printed on. These technical and aesthetic aspects of the bottle were important elements in its capacity to establish multiple associations in production and distribution chains, in marketing processes, and for consumers.

The emergence and the effects of the PET bottle as a mass object were also shaped by its capacity to displace other packaging materials in the beverage industry. This process of material substitution was a significant force in the ongoing actualization of PET as a market device. As a new form of packaging, the PET bottle had to be qualified in relation to existing packaging forms and the liquids they contained. A key element in this process was how the PET bottle interacted with the incumbent materials in beverage packaging—how its qualities became calculated as superior. Industry accounts of the rapid rise of PET position the plastic bottle in an almost revolutionary relation to glass. It is described as a substitution packaging for the glass bottle that rapidly pushed the latter aside to become a major competitor with the can, and then the market leader in beverage packaging. However, the interactions between and among glass, PET, and aluminum were somewhat more complex than a straightforward process of substitution might suggest. Initially, the characteristics of PET bottles and glass bottles were defined in terms of similarity. PET was hailed as having qualities that were equal to those of glass (NAPCOR n.d.). It was promoted as the first plastic to match the optical standards of glass and to achieve equivalent translucency and clarity. This made replacement of one material with another easier, as consumers' expectations of being able to see what they were drinking were not radically disrupted. Like glass, plastic revealed the contents of the bottle to the buyer, preventing unpleasant surprises: it satisfied curiosity (Cochoy 2012). However, unlike glass, the PET bottle was unbreakable. It had durability without fragility, and this was a crucial difference.

In relation to the aluminum can, the PET bottle was positioned as offering similar portability and mobility, but the bottles also suggested a range of new drinking possibilities. For a start, they could be resealed—unlike the spring-pull can—and this made them useful containers for constant sipping over time and space, rather than solely on-the-spot consumption. This was a container that accommodated and enabled the mobility of the drinker. These multiple negotiations with other packaging devices in the beverages industry were complex. They involved the dynamics of market positioning that was structured by relations of similarity and difference. However, as PET became the preferred and dominant packaging material, the PET bottle emerged as a market device in its own right. It had acquired the capacity to diminish the value of other packaging devices and to displace them.

In these patterns of emergent causation—making PET a mass object and rapidly displacing glass and aluminum cans in single-use packaging—it is possible to see how PET was qualified as a positive force, as a source of beverage market transformation and expansion. However, certain unpredicted effects emerged as the PET bottle was adopted more widely—specifically, growing amounts of plastic waste in urban waste management. While glass beverage bottles and cans had already established the practice of throwaway containers in recreational drinking, PET prompted new disposal practices. Discarding glass and cans often required a certain amount of care. Many, of course, entered waste streams, contributing to the massive growth in consumer waste after World War II as the use of disposable objects increased (Gandy 1994). However, the fact that these forms of packaging had been around for a long time meant that in many places, public bins, recycling, or container deposit systems had developed as alternatives to thoughtless discarding. Both glass and aluminum also presented good opportunities for recycling and economizing waste, as these materials could be reused in the making of new containers.

PET bottles disrupted these arrangements when they swamped the market, for the plastic recycling industry was largely undeveloped. PET could also only be *down*cycled because it was impossible to get the same optical clarity using discarded plastic. Each new bottle required raw, not recycled, materials. This absence of effective waste management systems enhanced the connotations of PET as *more* ephemeral and disposable than other packaging. As Freinkel (2011, 174) argues, "the introduction of light, unrefillable PET bottles helped seal the changeover to what the industry calls 'one-ways.'" This was a significant and unexpected effect, one that Freinkel regards as implicated in the decline of sustainable two-way waste management systems such as container deposit schemes.

In this way, the PET bottle helped inaugurate the concept of "nonreturnables" in the bottled beverage industry and the significant urban waste problems that resulted. This industry category didn't just disrupt two-way schemes for waste management, it also gave new and troubling meanings to the idea of convenience and disposability. The lack of any obvious pathway for the afterlife of the bottle either through recycling or container deposit schemes amplified the perception that plastic had only a present or immediate temporality. This inhibited consumer awareness about the possible futures of the object as solid waste. Putting the empty glass bottle or can in the recycling bin or saving it to collect a small recompense extends the temporality of the object after the moment of immediate consumption. These sociotechnical systems suggest to the consumer that the object has a future beyond waste, and that it has other possible values. The absence of this sociotechnical infrastructure when PET bottles took off fueled the idea of plastic as eternally available and eternally present. The affordances of PET as transparent, almost weightless, barely there—unlike the solidity of glass and cans—also seemed to make littering easier. In these ways, the PET bottle suggested cavalier waste practices. It enhanced the idea of disposability as fleeting utility. So, while PET was rapidly adopted as a replacement for glass, its material performance generated new discarding practices that became increasingly problematic. This unexpected evolution of the object into a one-way thing with serious effects on waste management shows how the PET bottle was an "unfinished object," something whose object status was continually evolving in relation to the connections and interactions in which it became enmeshed.

These are some of the patterns of emergent causation that reveal how the PET bottle became active as a market device in the packaging industry—how its diverse forms of agency evolved, generating the capacity to produce significant effects. These effects were shaped by various dynamics, such as the logics of packaging production and beverage markets, which enrolled the PET bottle in specific ways. At the same time, the materiality of the bottle interacted with these dynamics and made its own suggestions, from integrated production and filling processes to massive increases in plastic waste. Emergent causation challenges the idea of a simple linear process of industry application or revolutionary material substitution. It presents economic assemblages as patterned by diverse and distributed forms of agency beyond corporate and human intentionality. Through various processes of discipline, massification, and interference with other packaging processes, the *economic expressivity* of the PET bottle emerged and the bottle's capacity as a market device was enabled. But so too were many other disturbing capacities or shadow realities.

The point is that the PET bottle did not enter into the beverage industry with a fixed identity that caused certain effects. Nor was it the passive instrument of inexorable economic imperatives or corporate intentions. Rather, it became a participant in already existing markets and, through its interactions and relations with other devices, helped reconfigure and extend these markets. The PET bottle forged a range of important alliances with continuous-flow oil refining, bottling plants, multinational beverage companies, and consumers, and as it did so it gained strength to the point that it became capable of profoundly rematerializing beverage packaging and the ways in which people drank and discarded.

Forming Alliances between Plastic and Water

This account of the diverse processes whereby the PET bottle was enrolled as a market device is necessarily broad. Its value is to show how this form of packaging rapidly came to matter, how it was qualified in relation to other forms of packaging, and how its evolving forms of agency in beverage markets emerged. It is impossible to understand the implications of packaging water with this container without this general account because it shows how small mundane market things can drive major transformations, as Franck Cochoy (2007, 120) argues: "Packaging changes the product, the consumer and the producer all at once." How, then, did the PET bottle specifically change water and the ways in which consumers apprehended and valued it? In what ways was the performativity of the package a crucial sociotechnical element in the organization of new economic actions around water?

As we have discussed, packaging water in PET bottles was an event not simply because it caused significant transformations in beverage markets, everyday drinking habits, and more but because it also rapidly became controversial and a political issue. Although bottled water had been around long before the rise of the PET bottle, it was generally distributed through boutique markets that sold water from mineral springs in glass bottles (Marty 2006). The glass bottle often enhanced the qualification of the water as from a distinctive source, with special chemical components or as something to accompany the dining experience; hence the classifications of "mineral water" or "table water." Water packaged in PET bottles represented a very different set of alliances between the packaging material and the contents, and a very different economy of qualities for water. As the activist critiques of the industry show, the rise of the PET bottle was fundamentally implicated in enabling the development of an immediate consumption market in single-serve drinking water: a mass plastic colluded with an essential liquid to make it into a mass commodity. The issue is, exactly how? How did this particular packaging material solicit the product? How did PET bottles and water establish relations of reciprocal influence that worked to mutually transform and enhance their qualities?

The answers to these questions, especially in trade industry journals, usually point to the exceptional functionality of the bottle in making new forms of water containerization and consumption possible. PET is a favorite in the beverage industry because of how it has been able to contain a readily available liquid and produce new desires and uses for it. Marketing and branding are also recognized as playing important roles, but the consensus is that this plastic bottle is remarkable for its practical capacity to build markets in water in "industrialized countries ... despite the good quality of drinking water available directly from the tap" ("PET Remains a Favourite" 2008). This literal explanation, however, does not account for the complexity of exactly how water has been rendered economic and the immense efforts, processes, and devices that have gone into this relatively recent development. Nor does it help explain the ways in which packaging became a crucial element in making water calculable and in building consumer attachments to these calculations. To understand how the PET bottle became a constitutive feature of bottled water markets, it is necessary to investigate the intricate and intimate connections it established with both the contents and the consumer. The previous section outlined how the bottle was incorporated into existing packaging and beverage production, but here our focus is on the changes this container surreptitiously introduced into consumer perceptions of water, and the new dispositions it suggested in relation to water. Obviously, many other elements were involved in making water calculable, and we explore them in the next two chapters, but here the aim is to understand the ways in which this mundane container inaugurated a more mediated relationship to water, how it was implicated in qualifying it in specific ways, and how it helped build new attachments and understandings of the place of this liquid in daily life.

Contemporary accounts of the rise of packaging (see Bowlby 2000; Cochoy and Grandclément-Chaffy 2005; Hine 1995; Twede 2012) offer important analyses of the role of packaging in reordering relations between products and consumers during the twentieth century. They show how packaged goods inaugurated a more mediated relationship to products, one in which consumers had to rely on indirect, written or visual information to access knowledge of what they were buying. Direct physical encounters with the product were replaced by apprehension of the package, which contained information about the physical, scientific, or cultural dimensions of what was being purchased—information that was previously unknown or inaccessible. The growth of packaging made it possible (and, in fact, necessary) to invent and highlight product differences, and this necessity fueled the expansion of marketing and brand strategies, the package providing a crucial surface for establishing the symbolic qualities of the product and other forms of brand development. As packaging became normalized in market assemblages, consumption became impossible without it. This was not simply for technical and pragmatic reasons but because shoppers expected packages: packaging had become an integral part of the sociomaterial meanings and practice of shopping.

In this context of a shopping experience and retail world dominated by packaging, things that are *un*packaged stand out. As Cochoy and Grandclément-Chaffy (2005) argue, they appear "naked." How is it possible, then, to understand the appeal of something, or the ways in which it is located in market relations, without the material and immaterial networks that packaging realizes? In relation to water, these questions are critical because, in its service or reticulated form, it is naked or unpackaged. It might be piped and networked, but these mediating devices are largely invisible to the consumer: the water flows in seemingly direct response to consumer demand. While consumers might sense that they are accessing a vast sociotechnical capacity or public infrastructure when they turn on the tap, the water still has an immediacy, a directness, that is captured in the seemingly never-ending flow that the consumer controls. In this way, reticulated water generates a specific ontological reality that is dramatically transformed when water is accessed from a container. These transformations occur in many registers: materially and semiotically, economically and politically. Although they are interconnected, we consider them in turn before examining how the PET bottle puts water into a new series of relations and helps make it a market thing.

As we saw in the previous section, the specific material affordances of the PET bottle played a key role in industry take-up and consumer attachments. However, when it came to containerizing water, the bottle's light weight, extreme clarity, and shatterproof qualities were significantly amplified, both aesthetically and technically. The specific qualities of this plastic seemed almost to mimic and enhance the biophysical qualities of water. PET was exceptionally transparent, with an optical density very similar to water's. Light passed through it in much the same way as it passed through water, generating an effect of fluidity and shimmering purity. Water seemed *shinier* in bottles. This made the perception of transparency more intense in the sense that, while the bottle contained the water, it also simultaneously seemed to make it appear almost unpackaged (see figure 1.2). The functional capacities of the package—its role in holding or containing the water—merged with its material-semiotic power to both enhance and imitate the qualities of the actual product (Bowlby 2000, 102).

Figure 1.2
The transparency of plastic and water.
Source: Gay Hawkins.

The bottle was also very light and thin, which meant that holding it generated a more tangible contact with the contents, despite its functionality as a barrier and container. These distinct sociomaterial mediations in which the bottle and the water formed potent affective alliances had the effect of the package *becoming* the water, almost in a process of transubstantiation. In design terms, this was the triumph of packaging: not only did the first moment of apprehension of the bottle on the shelf appear to the consumer to be an almost direct encounter with water but the sense of being able to see everything, of the bottle's complete transparency, also reassured consumers that what they saw was what they got: water that was shinier and purer than any that flowed from a tap: "From the consumer standpoint package and product are one and the same" (Le Guen n.d.). All these effects were amplified by the work of the brand, which we explore in the next chapter, but even before branding and product information, the packaging had achieved an enormous amount in terms of requalifying water and making it appear very different from the prosaic flows of the tap.

Economically, the package inserted water into beverage markets and helped diversify those markets. These economic effects were intimately connected to the material-semiotic ones described above, but they were also distinct. The container established a whole new force field of relations for water that helped its framing as a market thing. Of most significance was the way in which the container literally and figuratively detached water from its everyday reticulated settings and offered itself as a new water *source*. In her essay on container technologies, Zoë Sofia (2000) argues that the container becomes the source of what it holds or preserves. It isn't simply performing a static function because containing is an action in itself; the dynamic capacity of the container is to both hold and *re*-source. With this argument, it is possible to see how the PET bottle was a powerful sociotechnical device for economically reframing and re-sourcing water. In the same way that the bottle was designed to be easily handled, with its curvaceous form adapted to the hand grip and to portability, it also allowed consumers to feel they had access to their own personal water supply. The bottle reframed water as an individual resource, helping to transform an environmental and often public resource into a personal possession. While the work of the brand often elaborated extensive information about the unique "natural" water source in order to distinguish it from tap water (see figure 1.3), the dynamic material capacity of the container did much to detach water from its existing sociotechnical framings. It *re*-sourced water in ways that were both practical and also fundamental to the organization of markets.

The bottle reorganized relations to water on many levels: it requalified water and helped create market forms of worth for it where they were generally not present, especially in places where safe public infrastructure was the norm. This occurred through the logic of re-sourcing and of market standardization, such as the use of volumetric measures and individual or single serves. Although many other elements were involved in the process of economizing water, the bottle was critical to rendering it in a serialized form and making it perform like a commodity. And whereas these processes were familiar to drinkers of sweetened beverages, they were relatively new in relation to water, which until the late 1980s did not have a big presence in mass beverages markets. In assessing the effects of these processes of reframing water and promoting new forms of consumption, the issue is not how a product undermines or straightforwardly displaces a service. Rather, it is how the bottle multiplied the possibilities for how water could be "sourced," related to, and valued, and how these new market ontologies interacted with others.

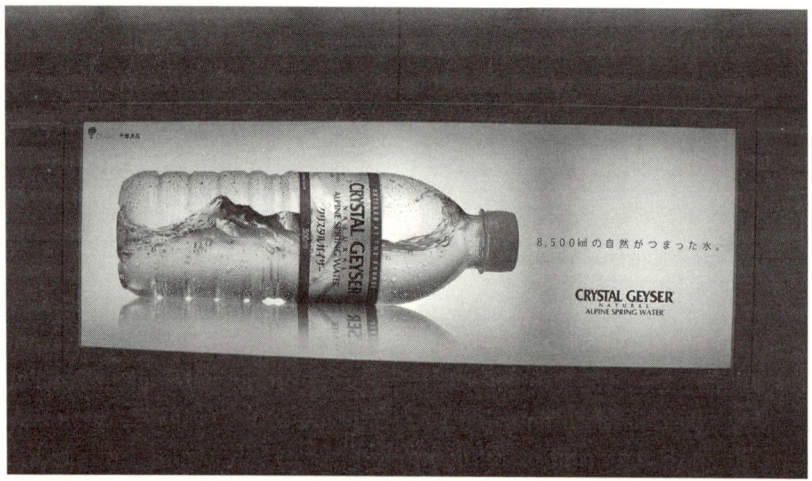

Figure 1.3
The bottle re-sourcing the water.
Source: Warwick Pearse.

Conclusion

Two stories have been told here, the history of the invention of the PET bottle and how this new form of plastic packaging became a powerful actant in the beverages industry; and the story of how bottles and water established relations of reciprocal influence that had the capacity to generate significant economic and biopolitical effects. The performative agency of the bottle of water at loose in the world is many-sided and the focus of later chapters. In this chapter we have sought to explain why the invention of PET packaging had the effect of rematerializing the beverage industry and how the subsequent filling of these bottles with water was an event, a development that triggered the proliferation of new relations, meanings, and values of water.

2 From Containers to Contents: Branding Water

The invention of the PET bottle was central to the phenomenal growth of bottled water markets. Of course, water had been containerized and sold long before PET was developed. The issue is not that PET *invented* bottled water; rather, as a new form of plastic packaging and sociotechnical device in the beverage industry, PET played a key role in helping to assemble mass markets in single-serve water and to reconfigure how water was valued. The plastic bottle, in association with many other elements, was the source of generative effects; it acquired the power to make differences. But what of the contents—what of the actual liquid that the package contained? How did the water in plastic bottles acquire new characteristics that distinguished it from other drinking water, especially the ordinary flows from household taps? The ability to take a swig from a bottle of water is the outcome of dense alliances between and among material, technical, sociocultural, and economic processes that are vastly different from those involved in turning on a tap. Tap water is essential to many of these bottling processes, from rinsing the freshly blown PET containers to (in many instances) filling the bottles, but when the water in a bottle appears as a product, these messy industrialized production processes are disavowed. Instead, we apprehend water that is swarming with new qualities and that is implicitly differentiated from the ubiquitous liquid flowing from taps.

One of the key processes that help organize these new qualities is branding, and the diffuse set of practices that surround it—from marketing to labeling information to consumer research. While packaging design is very closely implicated in branding strategies, and these processes are often developed in tandem, here our concern is with the distinct operations of branding in the framing and requalification of water. Of particular interest is the historical emergence of branded water in the nineteenth century, and the ways in which markets in boutique mineral waters developed strategies for communicating new values for water. This empirical and historical focus is useful for understanding the contemporary role of brands in mass bottled water markets. The point is not that boutique mineral waters caused mass markets through some logic of inexorable market expansion but rather that branding has been central to assembling diverse markets in water for a long time: the issue is how, and with what effects? In what ways do contemporary packaged water brands draw on the cultural framings established by earlier boutique water markets? Our claim is that the brand is not simply a semiotic and technical device central to market development but that in recent years it has prompted significant transformations in the wider valuations of all drinking water. The brand inaugurated new ways of classifying and relating to water; it generated diverse effects, and the scope of these effects—both historically and in the present—is our focus here. Like the plastic PET bottle, the brand appears to be a potent element in the event of bottled water. It is an object with inventive and extensive consequences (Lury 2004, 151).

The Work of the Brand: Bottles and Taps

A quick glance into the beverage refrigerator in any supermarket reveals that water in a bottle is branded water. Some of the brands have global recognition and represent massive transnational networks of distribution and consumption. Others are local and unknown beyond the bounds of very restricted spaces of circulation, but the brand persists. Stamped on the transparent surface of the PET bottle is an image or logo that doesn't simply qualify the water but also situates it in a complex network of relations. Labeling information surrounding the logo might detail the place where the water has come from, or its biochemical components, or health advice on how much consumers should drink every day. In other words, brands do far more than stand for or represent the product. They are, as Lury (2004) argues, complex ontological objects and translation mediums that shape relations between products and consumers and create fields for various social actions, not simply exchange. The issue is how the brand's commercial regime of value interacts with other nonmarket values for water. In what ways do market forms of worth create new fields of action that are not simply economic but also cultural and political?

In posing these questions, our aim in this chapter is to understand branding as a critical device in the emergence and coordination of new markets in water *and* new biopolitical dynamics around water provision. Of particular concern is the way in which branded bottled water interacts with unbranded tap water—how water as a singularized product is differentiated from, but also implicitly related to, other forms of water. According to Michel Callon, Cécile Méadel, and Vololna Rabeharisoa (2002, 201), central to all markets is an "economy of qualities" that is produced through the processes of qualifying and positioning goods: "Through construction, a product is always singular and similar to other products because it is immersed in a space of qualities that makes comparisons possible." This argument is clear in relation to competition between different bottled waters—between, for example, Evian or Fiji Water—but what of the competition between Évian and tap water? How are competitive relations between water framed as a product and water framed as a service organized, and what are the effects? Does tap water inhabit the same "space of qualities" that markets in bottled water inhabit?

At first glance, the answer appears to be no. Delivery of water via governmental infrastructure is organized along very different lines. It is most often provided by a monopoly; the user is charged on the basis of access to a connection and metered rather than volumetric use; water quality is regulated by public health authorities; the water is usually sourced from surface rather than from underground supplies; and it is delivered by a vast sociotechnical network. This complex arrangement assembles a very different set of qualities for and relations with water. The continuous flow from the tap does not generate a multiplicity of related products from which the user must choose; rather, it constitutes a service relation, a "making available" through access to a sociotechnical capacity. While there is no question that urban water supplies and infrastructure are increasingly subject to various market dynamics, these market forms are vastly different from those shaping bottles, and they produce very different water.

Despite this apparent difference, however, the water in bottles and the water issuing from taps are fundamentally connected, and brands are central to this relation. If brands are a locus of value and if all valuing is relational (Frow 1995), then other waters provided by other mechanisms are implicated in the new space of qualities that the emergence of branded water has created, *whether they like it or not*. Brands have very effectively requalified water by introducing new calculations into its consumption. These calculations are not only quantitative; they are also qualitative and affective (Lury 2004, 7). Drinking Cloud Juice or Fiji Water allows you to access water that is "untouched." Drinking Dasani protects you from the risk of unknown impurities lurking in other waters. The work of the brand is to reveal to the consumer new meanings and values for water, to intensify its qualitative possibilities. Implicit in these framings is a differentiation from tap water. If Fiji Water is "untouched," then, by implication, other waters must be suspect because of their contact with infrastructural sources of collection and piped distribution. If Dasani is "safe," then other waters are unsafe. Not only have brands requalified water by positioning it in relations of differentiation from taps, they also often implicitly suggest the substitution of tap water with bottled water as a less risky or healthier option. As the authority of the brand is mobilized in revaluing or qualifying water, water that is unbranded—that is anonymously delivered through a network—inevitably appears inferior, because to value some things is also to devalue other things (Frow 1995; Smith 1988). Branding is potent and effective because it brings things into relation and differentiates them at the same time. This dynamic plays out in different ways in different urban water contexts, but the general point remains: the work of the brand in making new water markets inevitably implicates other forms of water. By introducing new calculations that are meaningful only in relations of differentiation, requalification, and, often, substitution, branded water mediates the wider supply and valuation of other drinking waters.

This observation, that branded bottled water has implications for other forms of supply, is not new. In many current analyses of bottled water, brands are often the primary focus (see Wilk 2006). In investigating the relatively recent rise of mass markets in single-serve water bottles, this work recognizes the historical links between the intensification of corporate practices—including branding—since the 1980s and the growing privatization of more and more economic and social functions. However, while brands are recognized as important, the tendency is to simply equate commodification or privatization with branding, or to analyze the brand as a cultural phenomenon central to deceiving consumers by duping them into paying a massively inflated price for something that is relatively cheap.[1]

From our research into bottled water, the role of the brand requires more considered attention, focused on exactly how it participates in assembling new forms of economic activity and value around water. Brands are not only a significant cultural form, they are also an increasingly potent modality of economic power (Lury 2004, 10); the issue is how this power is realized, and with what effects. Recent work on the history and dynamics of branding (Arvidsson 2006; Lury 2004; Moor 2007) shows that in the last thirty years, brands have become central to business strategy and to relations between consumers and companies. Branding strategies are not only being applied to more and more institutions, services, and things, but managing the brand, and developing and exploiting its informational and wealth-generating potential, are now central to the organization of diverse economic functions. For Lury (2009, 72), these developments are evidence of brands' growing multidimensionality and the ways in which their functions and effects have become more complex and powerful. The recent rise of mass markets in bottled water historically parallels these significant changes in branding, and prompts the question of whether mass markets in single-serve water could have emerged without the rise of the branded economy? In what ways has the growth of branding provided important techniques and resources for transforming how water is apprehended and for multiplying water's qualities? There is no question that branded water produces complex new relations between and among consumers, water sources, and corporations. The challenge is to investigate the precise ways in which brands have come to function as a new modality of economic power in relation to the biopolitical provision of water.

This focus on the power of brands means that in the case of water, they have to be understood as market *and* political devices. According to Lury, brands do more than help assemble economic processes; they are also thoroughly implicated in the emergence of new social and institutional orders that subsume "the calculation of symbolic (and social) capital with the calculation of economic capital" (Lury 2004, 10). Brands make the calculation of economic capital much more than quantitative; they generate fields for the production of qualitative difference and possibility. Their force is both actual and potential or virtual. Arvidsson (2006, 13–14) develops this argument in relation to the potential of the brand to reconfigure the conditions of life:

Brands are thus an example of capital socialized to the extent of transpiring the minute relations of everyday life, to the point of becoming the context for life.... The brand, like informational capital in general, works through the bio-political context of existence to subsume the most basic and fundamental qualities of human life.

Water is thoroughly implicated in sustaining the basic qualities of life, and in enabling various social bonds. The branding of *water* prompts concerns because it is a "non-substitutable resource" (Bakker 2003) that is often cast as too special to be left to markets. The issue is the ways in which the interactions between branding and water have prompted new social imaginaries whereby water that is branded becomes seen as the best context for supporting life. This presumes that the work or performativity of the brand has complex and variable political dimensions that are not unilateral. Explaining the rise of branded water as the delegation of water provision to markets, or as privatization or relentless neoliberalism, does not acknowledge the complexities of the brand as a political device. In contrast, our focus in the rest of this chapter is on the precise strategies that branding water entails, and the ways in which these strategies enable new calculations about the relations between water and life—calculations that make trouble for the idea that water should be exempt from market dynamics.

The Case of Evian[2]

In the face of so many bottled water brands and so many different waters, issues of emergence, politics, and market assemblage are impossible to generalize. Instead, in this section we investigate how one market and brand in bottled water were assembled. This makes it possible to understand historical processes in the singularization and qualification of water, the ongoing invention and performativity of the brand, and the ways in which its potency as a market and political device has developed—especially in relation to the recent rise of mass markets in branded bottled water. Beyond this specificity, however, are patterns of general modality that are common to the function of many brands. This is not to argue that one case reveals everything but that the structural logic of branding is repeated across numerous bottled water markets.

Our example is Evian, one of the most recognized brands of still mineral water in the world. Evian, like Perrier, Vittel, and Badoit, is part of a prestigious cluster of mineral waters that makes France the sixth biggest market in the world for bottled water (Clarke 2007, 40). Evian has a long history as a branded water, being bottled and distributed from an underground source at Évian-les-Bains on the edge of Lake Geneva since the early nineteenth century. From this beginning, Evian has grown from a niche commodity to become one of the most popular table waters in France, helping to establish bottled water as a standard fixture at many French meals from the 1960s on (Marty 2006). In 1970 the Evian mineral water company was bought by BSN, a leading glass-packaging company that was diversifying from containers to contents. BSN was shifting its priorities to the food industry, having realized that, with the rise of cheap disposable plastics, glass was on the decline (Départment Médias Études et Communication Danone 2006). In 1994, BSN Groupe became known as Danone, now a major global food and beverage company. At this stage, Evian was the flagship brand for the soft drinks division, and it has become even more important as Danone has grown. As well as being central to the growth of mass markets in bottled water in France, from the 1970s Evian was increasingly exported to various international markets as a boutique mineral water. Along with Perrier, it has been identified as one of the key brands that established niche markets for bottled mineral water in the United States in the 1970s, prefiguring the rise of mass markets from the 1990s onward (Royte 2008).

Evian is a valuable example for several reasons. First, it can be considered a "heritage brand" (Lury 2009), having been associated with containerized water since 1870, when the statutes of the company Société Anonyme des Eaux Minérales d'Évian-les-Bains were approved. In its earliest uses, this water was presented as therapeutic, part of a wider culture of water treatments in vogue at the time.[3] Understanding the rise of drinking mineral water as an elite health practice is helpful in illuminating the wider historical context of contemporary bottled water markets. It makes it possible to see how earlier niche markets in water were assembled, and how these processes of assemblage might or might not prefigure current mass markets. Second, the early history of Évian water offers valuable insights into how water as a biophysical resource becomes a product, how the qualities of the water are stabilized, and how particular consumption practices are organized. Finally, this brand is significant because, like many heritage brands, it has changed and adapted over time. It has retained durable values associated with the "prestige waters" category of the beverage industry, but it has also been able to requalify and reposition the water in numerous ways in response to the rise of mass markets and the proliferation of a multitude of new branded waters in the late twentieth century.

In tracing the emergence of Évian water as a branded product, our focus is not on patterns of inexorable commercial expansion. This assumes that a market's diversification and differentiation adhere to a fixed linear logic. This is not the case: markets are delicate mechanisms, continually engaged in reflexive activities with various actors establishing new and often unexpected rules for the game (Callon, Méadel, and Rabeharisoa 2002). Since the late nineteenth century, markets in Évian water have constantly changed, not simply because of competition but because of their reflexive organization—that is, the ways in which they generate multidimensional relations that feed back into market processes and shape them in predictable and unpredictable ways. The changing role and performativity of the brand have been central in these processes. Lury's account of the brand captures this complexity well:

> brands are devices for the reflexive organization of a set of multi-dimensional relations between products and services, subject to statistical testing and the rapidly changing processes of mediation, stylization and practices of commercial calculation. But the brand as assemblage is not "simply" a social construction since the brand plays a part in the production of itself (and other things including products and markets). Nor is the reflexivity at work in the coordination of markets by brands representational, in contrast to, for example, the reflexive coordination of price. *The brand—as it is put to work—is a way not of representing but of modelling markets in many dimensions.* (Lury 2009, 78, emphasis added)

How, then, has the Evian brand been involved in modeling markets in many dimensions? In what ways does it function as the locus of shifting calculations and valuations of water? And how do the biophysical and material qualities of the water participate in or resist processes of market assemblage and branding?

To answer these questions, it is necessary to begin with an account of how water from a spring in the town of Évian became singularized and subject to specific processes of economization—how the water emerged as a market thing. Narratives of discovery of the miraculous curative powers of Évian water dominate existing accounts. These narratives belie the centuries-long history of water from springs in this region being used to supply domestic needs. This water was essential for the maintenance of populations, agriculture, and the local economy. In this long history, the water wasn't medicalized or exchanged as "mineral water"; it was a diverse biopolitical resource for supporting life. Long before the water from springs in Évian was bottled, it was enrolled in numerous associations with people, things, values, and technologies for capturing and circulating it. It wasn't natural, nor was it culturally constructed: it was a "quasi-object", the outcome of multiple interactions that were integral to the production of the social.[4] The processes of appropriating this water for new functions and exchange were complex and multiple. Making it into a product didn't simply privatize or alienate the water as if it were an original untouched external nature, it put it into different relations and associations that made it into a new quasi-object, that made it into a new water.

Making Water Calculable

In the standard histories of Évian water, the special characteristics or qualities of the water are assumed to be intrinsic. As you wander through the Evian Museum at the bottling factory in nearby Amphion or read the tourist brochures at the Évian Information Centre in the town center, you encounter a recurring origins story about how the spring water was "discovered" by the sickly Marquis de Lessert. It goes like this. In 1789 the marquis was visiting the region seeking out curative waters for his kidney troubles. He tried the water at Amphion, but it made no difference. He then went to nearby Évian on the shores of Lake Geneva and tried the water from St. Catherine's spring in Monsieur Cachat's garden. He began to feel better, and news of the powers of this miraculous water rapidly spread. Soon after this, doctors began prescribing the water, and Monsieur Cachat had to fence off his spring to protect it. In 1826 a Geneva businessman founded the first company to bottle and exploit the spring water, and at the same time a hydropathic establishment was built near the Cachat pump room for public bathing and water cures. In 1870 the newly formed company Société Anonyme des Eaux Minérales d'Évian-les-Bains began drilling for and purchasing other springs in the area. It also invested in hotels, casinos, and more bathing pavilions, helping to make Évian into a highly fashionable resort town for wealthy visitors seeking to take the waters. These people were known as *curistes*. Apart from dispensing the spring water in health treatments and spas, the company also bottled the water from the Cachat spring, labeled it "Source Cachat," and distributed it in big glass *bonbonnes* packed in straw to nearby Geneva and then Paris (see figure 2.1).[5]

This origins story obviously has promotional functions. In both the information center and the corporate museum, it makes marketing look like history. What is more interesting is the way in which it represents the relationships between the water, the place, the company, and the consumer.

Figure 2.1
Dispensing water as a therapeutic practice, c. late 1800s.

In the story of the "discovery" of Évian spring water's therapeutic qualities, the company is represented as a neutral mediator that simply makes the remarkable intrinsic qualities of the water available to the consumer. As a form of marketing, this narrative turns water into a story-laden substance, and this contributes to the ongoing stabilization of the product. It is part of the myriad processes whereby the qualities of the water are objectified. However, in seeking to account for how water from springs in Évian came into being as a branded product, it is necessary to write *against* this account, to analyze how the qualities of the water at Évian were not so much discovered as invented through various techniques of qualification and market assemblage.

According to Callon, Méadel, and Rabeharisoa (2002, 199), the qualities that are attributed to a product are both intrinsic *and* extrinsic, and emerge through various "qualification trials." Intrinsic qualities are the diverse material or physical characteristics that are present. Extrinsic qualities are those that emerge and are shaped by the specific devices and measures used in qualification trials, from regulations to techniques of evaluation designed to fix the characteristics of the product and stabilize its qualities in order to make it calculable: "All quality is obtained at the end of a process of qualification, and all qualification aims to establish a constellation of characteristics, stabilized at least for a while, which are attached to the product and transform it temporarily into a tradable good in the market" (199).

In the case of water from local springs in Évian, various techniques and qualification trials were deployed. These trials interacted with the intrinsic characteristics of the water and shaped them. In this sense, they were performative: the attributes of the product emerged in tandem with the specific means for testing and using it. After the claims made by the Marquis de Lessert about his miracle cure in 1789, various doctors conducted research on the treatment effects of water from Évian and published the results in medical journals specializing in water therapies. Other forms of research were also carried out, aimed at identifying the exact mineral components of the water and the unique biochemical and pharmacological qualities of each spring in the area.[6] This research was part of the growth of the emerging discipline of analytical chemistry and wider developments in the classification of springs across France. In the public baths and spa health centers that proliferated in the town from the early nineteenth century, the water was prescribed for numerous ailments, particularly kidney and urinary tract problems. It was often taken in doses administered by health workers of different persuasions, from doctors to white-coated spa attendants. For certain conditions, such as arthritis or rheumatism, it was recommended that the *curiste* bathe in the water. Gradually, strictly therapeutic practices were commingled with social and leisure pursuits, in which drinking the water became a form of conspicuous health consumption for elites. After the expansion of the French rail system in the mid-nineteenth century, Évian-les-Bains became more accessible to affluent tourists, and "taking the waters" became an excuse for bourgeois social visibility (see figure 2.2) (Coley 1990; Mackaman 1998; Marty 2006; Porter 1990).

All these activities—from chemical analysis, to scientific publications, to prescribing, to the building of spas, to bottling, distributing, and more—detached the water from its existing associations and put it into new relations. The effect of these diverse practices, qualification trials, and sociotechnical devices was to enact the medical *capacities* of the water, and to configure a market around these capacities. In these processes and new arrangements, the water was acquiring distinct forms of agency: therapeutic agency. It was

Figure 2.2
Drinking spring water at Évian as a form of bourgeois social visibility, c. 1914–1915.
Source: City of Évian Cultural Services.

coming into being as a different water, constituted by the new associations and usages it was thoroughly implicated in shaping and being shaped by. This is not to claim that the water was socially constructed, that the cultural and the economic obliterated the natural, but rather that processes of economization generated a relational materiality between the water and emergent market arrangements in which the water, with its specific biophysical properties, became an active participant. The demands of the market were to make certain aspects of the water present and calculable through interaction with qualification practices, from the authorization of science, to bottling, to water therapy. In this way, the earliest forms of market assemblage were not prompted by the intrinsic therapeutic capacities of the water but emerged as the markets were configured. As DeLanda (2006, 11) argues, the properties of elements involved in an assemblage—market or otherwise—do not cause relations. Properties only become expressive and potent with reference to their interactions with the properties of other entities.

The Political Mediation of the Market: The Role of the French State

However, probably the most crucial and definitive qualification of the water came from the French state. In a period of proliferating mineral water markets and spas in the nineteenth century, and amid debate about the validity of the health claims of various waters, the water from Évian could not be fully exploited as a commercial asset without its specific qualities being stabilized. The first decree governing mineral waters was issued in 1781. In 1856 a major law was introduced enabling the French state, after careful testing, to declare a water source or spring *d'interêt public*, meaning that the waters were considered beneficial to population health (Green and Green 1985, 15). This classification was applied to the Évian spring in 1878 when water from the Cachat spring won an award at the World Fair. This had an immediate impact on the growth of people seeking to both drink and take the waters at Évian. It also meant that the water could be bottled and sold commercially. Its health and mineral qualities were deemed such that the water could be safely drunk without medical supervision.

This nineteenth-century regulatory framework for mineral waters was administered by the Ministry of Health, the Academy of Medicine, and the Ministry of Mines. It had eight very strict rules:

• The mineral water had to be recognized as therapeutic by the state.

• The mineral water had to come from an underground reservoir alone, and had to emerge with constant temperature and composition.

• The mineral water could not be treated in any way (although filtering to remove iron was permitted in special circumstances).

• The mineral water had to be bottled at or immediately adjacent to the source.

• Not only did strict hygiene have to be observed but the entire operation was subject to constant checks by the Ministry of Health. The Pasteur Institute analyzed the water every two months.

• The Ministry of Mines established a maximum daily flow that could be tapped to ensure that the water table was not infringed on.

• Other drinks (such as soft drinks) could not be bottled at the same plant.

• A protected perimeter, usually six or eight kilometers in radius, was established around the catchment area, and no underground work of any kind could be undertaken without a special study by the Ministry of Mines. (Green and Green 1985, 16–17)

Receiving this authorization meant that water from Évian could be clearly distinguished from other mineral waters in a classificatory regime established by the state. Mineral waters that had the *d'interêt public* appellation were seen as having stable chemical and microbiological qualities that were beneficial to health; other waters could not make these claims (Marty 2006, 26).[7] This state-regulated standard meant that the distinctiveness of the water and its relationship to its source were both recognized and protected in law; the authority of the state guaranteed the qualities of the water. As a legal qualification trial, this regulation also shows how active the French state was in the nineteenth century in governing growing consumer food markets and in managing various forms of risk, from adulteration and fraud to spurious health claims (Atkins 2007, 980). Far from being an impediment to the supposedly natural evolution of markets, the state was critical in enabling their effective configuration. This legal ontology had significant power in constituting water from springs in Évian as a trusted and stable product, but so too did the water. The fact that it emerged from the spring with a consistent mineral composition meant that it was chemically stable and predictable, and therefore amenable to commodification. However, this material reality was emergent in relation to the regulation. The regulatory framework was a metrological device that translated the water into qualities that could be measured and made calculable. As a technique for measuring and protecting value, it was also fundamentally implicated in creating value.

The other effect of this legal regime was to constitute the water as "natural." In the classification of *d'interêt public*, a mix of medical, technical, geological, and political specifications was deployed in the explicit interests of protecting the water and the public. When the regulatory framework insisted that the "mineral water must not be treated in any way" or that "a protected perimeter of 6 to 8 kilometres must be established around the catchment," natural purity as a quality was not so much being protected as enacted. In this way, then, the new quasi-object of "natural mineral water" emerged out of myriad social, technical, and political qualifications that established the objective measures underpinning the emerging economy of qualities in water.

This brief history shows how Évian water became singularized and distinguished as a natural health product, and how various qualification trials made the water calculable. These nineteenth-century developments provided a crucial framework for the development of the brand. Initially, the water was labeled "Source Cachat"; in 1925 the label was changed to "Évian-Cachat"; in 1935 it became an official brand and was reduced to "Evian." Using the name of the town as the brand was effective in clearly differentiating this water from waters from other places. More important, it established an indexical relation between the brand and the objective source of the water. Lury explains indexical forms of branding in this way: "The relation of this sign to its object is indexical—that is, the sign denotes the object *through an existential connection to it*.... The Interpretant represents the sign as a sign of *actual existence*" (Lury 2004, 78, italics in original). This indexical relation meant that the brand Evian carried a trace of the origin of the water and functioned as an assurance of consistent standards, quality, and authenticity. It was a key strategy for provenancing the water, but it was meaningful only in relation to the legal ontologies that had already singularized it. The indexicality of the brand and the law amplified each other, and were powerful forces in framing the water *source* as the origin of its unique qualities. The law was a medium for translating and fixing the objective qualities of the water—for establishing the "facticity of the commodity," to use Atkins's (2007) term. The brand was a medium for representing and guaranteeing those qualities, and for communicating them across time and space.

The indexical dynamics in the earliest forms of the Evian brand were gradually transformed and elaborated throughout the twentieth century as the brand became not simply a sign of value but also a platform for assembling and generating it. These changes allowed the water to be revalued and repositioned in relation to significant changes in bottled water markets, specifically the recent expansion of fast-moving consumer goods (FMCG) markets worldwide. This is the reflexive capacity and flexibility of the brand: to model markets in many dimensions. This recent history is considered in the next section, where we investigate the rise of mass markets in bottled water and the proliferation of a multitude of brands. While Evian's markets grew steadily during the first half of the twentieth century, it was by no means a mass product. Until 1960, it was still sold only in pharmacies, and was still largely presented as a health product.[8] From the mid-1960s, corporate takeovers led to it being distributed in supermarkets and its qualities reconfigured as a superior table water: drinking it was pushed as a distinction strategy, not just a health practice. Although this had always been the case, the expansion of markets—particularly export markets—was predicated on promoting the water on the basis of its superior taste and its Frenchness, on building its value as a "prestige water" (Départment Médias Études et Communication Danone 2006; Marty 2006).

In the context of contemporary water markets, the history of Evian water is exceptional. The vast majority of bottled waters are not subject to strict regulations, and their claims to being "pure," "natural," or whatever are rarely authorized by legal standards beyond those already in place for the control of all beverages. On current Evian labels, drinkers are assured that this is water "bottled at source from the Cachat spring." This information doesn't just have an impact on pricing, allowing Evian to be placed at the top end of the market; it also signifies that drinking the water has more objective benefits as a health or distinction practice because its "natural" origins are actually guaranteed (see figure 2.3).

What is most instructive about the Evian example, however, is the way in which it highlights the complexities of making water calculable and the absolute contingency and ontological instability of "the natural source" in this process. The natural and pure status of Évian water was legally constituted and stabilized by the French state in the late nineteenth century. This political mediation remains fundamental to shaping and enabling the market because it places clear restrictions on the industrial production process, determining how the water can be accessed and bottled. In the case of other bottled waters, where standards are lax or completely absent, claims about the unique natural values and source of the water depend much more on the signifying force of the brand and other sociotechnical practices, from market positioning to packaging to distribution. This is not to say that the

Figure 2.3
Billboard in a car park in Évian town center, "Welcome to our factory."
Source: Gay Hawkins.

brand is minor in Evian markets—far from it. Rather, that brands are strategically enacted in different ways with different effects according to specific market assemblages. When it comes to water, singularizing what is already easily and cheaply available for many people is a challenge. The brand has to do a lot of work positioning and communicating special qualities for a very ordinary material.

Enacting New Qualities for Tap Water: The Example of Dasani

In moving now to an example of a contemporary branding strategy for bottled water, we consider how contemporary strategies have been implicated in making bottled water into a mass product, how they have contributed to the bewildering proliferation of waters over the past thirty years, and how this sense of an abundance of branded waters interacts with the emergence of concerns about water scarcity, the risks from consuming tap water, and wider threats to the public provision of water.

In contrast to Evian, with its long and documented history, many bottled water brands appear to come out of nowhere: they have no heritage. Although many brand strategies often reiterate discourses of health and natural purity established by the heritage brands—that is, they reflexively interact with the existing space of qualities that boutique waters have created—they also assemble very different cultures of water and consumer attachments. When one looks closely at how these brands work, it is difficult to claim that the rise of mass markets in water simply represents the democratization of an elite practice. This may be part of the story, but it is by no means the whole story. Rather, the mass markets indicate a diversification of the qualities of water as a product, and the ways in which branded bottles have positioned it in a multiplicity of new economic and affective relations. They indicate the significant work that brands do in assembling new values and fields of action for water and drinking.

As we have argued, the emergence of mass markets for bottled water parallels significant changes in corporate branding strategies, or what has been described as the "rise of the brand." This phrase alludes to developments in market practices and organization from the 1980s in the course of which the brand changed from being attached to a stand-alone product and began to develop and be deployed in multidimensional ways (Lury 2009, 72). This development of the brand involved strategies such as extending the product mix into related goods and services (brand diversification) and intensifying the communicational image of brands by attaching them to new activities, from sponsorship, to social causes, to employee loyalty programs. Positioning the brand in relation to a media-intensive culture also became as significant as positioning it in relation to similar products and to consumers (Arvidsson 2006). Underpinning these shifts were significant changes in marketing research that documented consumer practices and preferences and incorporated this knowledge into the ongoing requalification of products (Cochoy 1998; Lury 2009, 71). All these developments enabled brands to be strategically deployed in the expansion of existing markets or the creation of new ones. They were central to making mass products flexible and adaptive to constantly changing conditions, and to aggregating diverse consumer collectives or populations organized around any number of transient commonalities beyond standard market segments or demographics.[9]

The case of Dasani exemplifies many of these developments. The Coca-Cola Corporation launched this brand in the United States in 1999. At the time, Coca-Cola was distributing various international water brands, including Evian and several regional brands, but it had no house brand targeting a national market (though its rival, PepsiCo, had launched the house brand Aquafina). Announcing this new brand in an interview in the *Beverage Digest*, the CEO of Coca-Cola, Doug Ivester, explained the rationale for diversification this way:

Water is one of the fastest growing beverage categories ... it's a "tricky" subject. We have to be careful we don't replace high-margin soft drinks with low margin water.... We need a system-wide water. That's what Pepsi have with Aquafina. (Ivester 1998)

Like Aquafina, Dasani was designed for the lower end of the market, with Evian and other Danone waters that Coca-Cola distributed pitched to the top and middle. Dasani was promoted as a "purified water"—that is, tap water that had undergone reverse osmosis treatment. Coca-Cola's consumer research had shown that consumers in the United States were as happy with purified water as they were with spring water. Both these classifications seemed to be adequate for differentiating bottled water from tap water. The extra treatment of tap water was perceived as introducing new qualities into the mains supply, in this way producing a "purer" water. Before the company launched the brand, there was debate over whether to also add minerals. The minerals would have been distributed to bottlers as a concentrate, giving the brand a point of distinction from Aquafina and creating a purified and "mineral-enhanced" water, but the idea was eventually abandoned. The packaging did not involve a proprietary design but used an existing light blue PET model created for the sports drink Surge. This made production using existing bottle lines much easier and more cost-effective (*Beverage Digest* 1999).

Dasani was regarded as so open as a brand name that it could mean many things to many people. Like many other brand names, "Dasani" is a nonsense word that doesn't have any preexisting meaning but instead acquires a force through repetition that makes it into a "self-signifying proper name" (Frow 2002, 64). The parent brand, Coca-Cola, provided an important qualification of the water, connecting it to the authority of the world's largest beverage company. The product launch was done by direct mail, sampling, and 20 million redeemable coupons made available to working women and parents by way of publications such as *Reader's Digest*, *Shape*, *Newsweek*, and *Parents*. The initial marketing slogan, "life simplified," positioned Dasani as a way to relax and find refreshment while "on the go."

The Dasani brand also sponsored "Replenishment Zones," and was plied at heavy-traffic summer sporting events (Howard 1999a, 14). Initially, the distribution of Dasani focused on immediate consumption channels such as vending machines, petrol stations, and convenience stores rather than on supermarkets. The water was inserted into ordinary life *on the move*; it was available, waiting for the consumer and offering access when consumers were far from the household tap.

This placement of the water in locations that connected with consumer movements rather than with deliberate visits to shops, such as the weekly supermarket visit, instated new meanings for "convenience." As Shove (2003, 171) has observed, convenience is about the scheduling and coordination of people and objects in time and space. It supports the intensification of urban life by reducing the time tasks might take with the provision of things almost instantly. This logic was exemplified in the promotional text on the Dasani redeemable coupon: "Life doesn't have to be that complicated. Life takes a lot out of you. Dasani helps put it back in. A moment with Dasani helps replenish both mind and body" (Howard 1999b, 12). While water in the household takes no time to prepare, having access to it outside the house while on the move does. It requires planning and equipment. Bottles for sale at key points of consumer passage addressed this reality. They made having a drink of water something that could be inserted into other activities; they made the bottle connect with mobile consumer practices and schedules.

As the brand evolved, new target markets were identified and a range of different marketing strategies was implemented that situated the water beyond networks of convenience. In a 2001 campaign aimed at young single women, the bottle was presented as an essential accessory for a big night out. Young health-conscious urban women were the focus of a television and online "wellness campaign." The tagline was "Can't live without Dasani." Central to this strategy was an ad featuring a female Dasani drinker enjoying a frenetic *Sex and the City*–style night of New York clubbing, with a bottle of Dasani in her hand. Described as "breakthrough entertainment positioning" (Lipperts 2003, 22), this campaign was significant for the way in which it shifted the economy of qualities for Dasani water away from relief from urban stress and toward issues of excitement, entertainment, and sex. Looking after yourself, or "wellness," was still a central part of the message, but this practice was now extended to all areas of life. Water was

not something you might spontaneously reached for in a gesture of immediate consumption; it was something you now *carried with you everywhere,*
along with your mobile phone and wallet. This positioning of the bottle
reimagined convenience in significant ways. It located access to water in
the rituals of mobility, in the new practice of carrying one's own supply,
and in forms of self-care and personal regulation. Related branding strategies in this campaign involved a variety of sponsorship and promotional
events, such as the Dasani/Heatherette T-shirt Contest, run at the School of
Design in New York City in 2002, a key example of intensifying the communicational reach of the brand.[10]

This brief history of the emergence of the Dasani brand shows how it
was developed in relation to the wider dynamics of beverage markets in
the United States, particularly the success of PepsiCo's Aquafina and the
increasing singularization of water as a product extension in the fast-moving or immediate consumption sector from the mid-1990s. However, more
crucial are the ways in which this brand interacted with the wider space of
qualities for water during this period. Dasani is a valuable example because,
unlike Evian, the bottles are filled with tap water. Coca-Cola had to build
an economy of qualities around "purity" and convenience rather than a
special or unique source. Although Evian has also always made claims to
purity, these claims are connected to an unadulterated spring source and
the regulatory framework that protects it. Dasani, in contrast, had to establish different measures for purity by *requalif*ying tap water, or the readily
available mains supply. In many senses, the ordinary and readily available
source, tap water, had to be disavowed by the "treatment," and by distinct
production processes that changed public water into branded water. How,
then, did this particular market transform tap water, and in what ways were
these transformations or qualification trials successful? The answers to
these questions differ according to the location of Dasani's markets. In the
United States, Dasani did exceptionally well, rapidly becoming the second
largest national brand within a few years of its launch. The reach and power
of Coke's distribution networks were a critical part of this success. Americans, it seemed, were comfortable with buying modified or "enhanced" tap
water, perhaps because levels of trust in mains supplies were increasingly
under attack.[11] However, in 2004 the brand was launched in the United
Kingdom, and in this new national context the water faced a difficult
requalification trial.

In March 2004, a controversy erupted in the London press over the newly launched Dasani water. *BBC News*, the *Guardian*, the *Observer*, and the *Financial Times* all ran stories documenting the new product on the market as nothing more than purified tap water. These articles all focused on exposing Dasani's "big secret." Their common theme was that the product was a deliberate attempt to trick consumers into buying something that they thought was different from tap water, that it represented a 3,000 percent markup, and that re-treating and bottling tap water explicitly undermined the existing value of this water (see Garrett 2004; Lawrence 2004a). The Dasani PR response to this market contestation was to promote the complexity and sophistication of the purification process, to explain how a "new" or better water was produced. An industry representative quoted in a *BBC News* online report described the process as "four stage," beginning with three separate filters, then reverse osmosis using a technique "perfected by NASA to purify fluids on spacecraft," followed by the addition of minerals to "enhance the pure taste," and finally "ozone injections" to keep the water sterile (BBC 2004). Market contestation was turned into market opportunity, with the company using the media attention to explain and stabilize the qualities of the product. It is possible to see how, in this four-stage treatment and the relentless promotion of it, the extrinsic qualities of the product were coming into being. These various sociotechnical processes were not simply the production process, they were designed to differentiate the water from tap water and make it calculable. In response to the claim that this treatment implied tap water was impure, the Coca-Cola representative said, "We would never say tap water isn't drinkable. It's just that Dasani is as pure as water can get—there are different levels of purity" (BBC 2004). Contestation forced the company to make public what had always been implicit: the space of qualities that bottled water created through various processes of singularization and positioning was meaningful only *in relation* to tap water.

Soon after this controversy over the qualities of water and the distinctions between bottled and tap water hit the news headlines, Dasani faced another crisis with a major contamination scare. The process of adding calcium chloride containing bromide to improve taste and then pumping ozone through the water was found to oxidize the bromide and turn it into bromate, a carcinogen. Testing of Dasani water by the UK Food Standards Agency showed that it contained levels of bromate that did not present an immediate risk to the public but did breach legally permitted levels. After just five weeks on the UK market, Coca-Cola voluntarily withdrew all bottles of Dasani (Lawrence 2004b). Coming on top of the media exposé about the tap water source, this contamination scare turned the UK launch into a form of public entertainment. It also prompted vigorous defense of the value of the tap from numerous municipal suppliers. Dasani was being positioned as Coca-Cola's entry into the rapidly growing UK and European bottled water markets, and challenges to its water quality did massive damage to consumer trust in the brand.

This account of the development of Dasani shows how the brand was involved in ongoing qualification trials for the product and how it worked in relation to a range of other strategies, from marketing to packaging to distribution networks, in assembling a market. It also shows how the brand can be considered as an interface or medium of translation (Lury 2004, 49) in the way it manages the relations between the complex environments of production and circulation, on the one hand, and the everyday environments and consumption practices of consumers on the other. Lury's characterization of the brand as interface highlights how processes of translation and exchange between these environments are dynamic and two-way, but also not direct, symmetrical, or reversible (50). Although companies typically expend considerable effort on marketing and on figuring out how brands and consumption help form particular identities in consumers, what remains hidden in this exchange is the identity of those who produce the goods being consumed and the contexts of production. Dasani represents an important example of this dynamic. In turning water into a branded product, Coca-Cola's main effort was geared toward inserting the bottle into everyday practices and organizing an economy of qualities for it as convenient, mobile, and good for you: part of the (usually female) consumer's identity as healthy and vibrant. The primary focus of the interface was on the external environment of the product: where and how and why it should be used. But the brand as interface also had to reveal the other environments of the product, its production process and source. It had to continually establish the qualities of the water as different from tap water, as produced in different ways, as "enhanced." It had to turn water readily available as a service into water branded as a product.

This process of stabilizing the product is never complete. As the UK launch showed, Coca-Cola had to reflexively respond to a rapidly changing field in the face of high-profile attacks from the media and ordinary citizens on water quality. Rather than the brand functioning as a locus of value and trust, it emerged as a product of dubious practices that were basically turning good public water into bad packaged water. In this instance, the Dasani brand was not a guarantee of consistency or trusted origins but instead came to represent a form of unnecessary and hazardous commercial interference. Extra treatment processes and market forms of value were seen as disrupting and diminishing the public distribution and value of water. Attempts to qualify branded bottled water in relation to tap water, to position it as superior, had the opposite effect.

Conclusion

Evian and Dasani are brands with very different histories, operating in distinct markets. But they are also connected through the wider space of qualities surrounding water as a product. Dasani exploited the regimes of value for water that the premium heritage brands such as Evian had established throughout the twentieth century. These earlier markets and heritage brands had established distinctive framings for water that enabled it to be disentangled from other associations and made calculable. Although such brands were only one of many devices that went into making water a market thing, their references to purity, to the source, and to drinking water as a health practice were all important in suggesting how recent mass markets could be developed. But this reflexive interaction between old brands and new, this positioning in relations of similarity and differentiation, works both ways. It is not simply that new bottled water brands have borrowed the framings of established brands. These prestige waters have also had to adapt to a massive intensification of competition and new market framings. In the face of a proliferation of new water markets from the mid-1980s, Evian extended its market reach beyond the premium sectors, seeking to make its water more popular and an everyday commodity through campaigns that addressed the fitness market, the partying youth market, babies, and many more. Evian thus had to requalify its water in relation to an abundance of new waters and mass markets. This was about more than just the ongoing dynamics of market positioning and market share; it was also about how the rise of brands intensifies valuing regimes surrounding water products and increasingly comes to shape the system of relations between them—how brands have become central to structures of competition (Callon, Méadel, and Rabeharisoa 2002). The brand does not determine these relations or express linear and singular corporate intentions but contributes dynamically and reflexively to the production of diverse economic and cultural processes surrounding water.

A central claim in this chapter has been that brands have been crucial to making water calculable, and that many of these calculations are meaningful only in relation to tap water. In this way, we have argued that brands have to be considered market *and* political devices. What the histories of Evian and Dasani show—though in very different ways—is that brands allow exchange relations to infiltrate the most ordinary relations of everyday life. Drinking water is one of the fundamental contexts of life. In places where water is made available as a state-provided sociotechnical capacity, branded water may well exist on the fringes of this form of provision. It may be restricted to forms of immediate consumption or substitution of other beverages. In places where safe water is not available, branded water provided by corporations is often normalized as the source of safe supply for those who can afford it, and we explore those dynamics in part II of the book. Branded water acquires different values and biopolitical effects in different settings. The point is not to unilaterally condemn water as a branded product but to understand how brands acquire political capacities in particular settings.

One of the key ways in which brands do this is by strategically exploiting uncertainty in relation to the provision of safe water. In the earliest forms of market configuration in the region of Évian, this was not the case. Branding the local spring water did not generate political interference with other public modes of supply, as these modes were not yet in place. Most people in the region were already drinking local spring water. Évian water as a product was largely consumed by visiting bourgeois health tourists in special settings and in special ways. Branding was a way of detaching the water from ordinary local settings and uses and qualifying it as distinct for very particular consumers.

In contrast, many contemporary brands opportunistically present a new economy of qualities for water that reconfigures existing relations with it, that offers an alternative source of trust to that provided by the municipal water supply. This is the managed reflexivity of the brand—not as a unilateral source of domination and control but as a flexible economic modality that can implicitly interfere with noneconomic or public modes of value. Brands are central devices or actants in market assemblages; they have made new economic actions around water possible. The ways in which many of these brands become political, not just market, devices is by appealing to the fundamental sense of water as a basic necessity of life. The constant referencing of life, even while bottled water brands are qualifying it or turning life into lifestyle, is a distinct regime of value that seeks to associate water as life with the brand. This is how brands acquire biopolitical capacities, and how market valuations of life gain force and effectivity.

3 Frequent Sipping: Assembling the Subject of Hydration

As we look around us in our daily activities, we can observe how slavishly the exhortation is being followed. Everywhere, people are carrying bottles of water and taking frequent sips from them.
—Valtin 2002, R993

This chapter examines an important element in the formation of markets in bottled water: making up the thirsty consumer. This can be understood as a process of creating attachments between people and bottled water—a process that involves much more than glossy advertising (though advertising is undoubtedly important). It involves tapping into the dispositional tendencies and practices through which people make themselves into subjects. In the case of bottled water, this process has entailed mobilizing discourses of health and personal performance that were already enjoying immense popularity from the 1980s on. More specifically, it has involved an appeal to biomedical languages and in particular to scientifically presented concepts of hydration, or the practice of drinking sufficient water.

If PET plastic and branding practices have enabled new qualities and practices to converge around water, what pulls these together is the thirsty subject: the subject who is led to drink water in particular ways. Obviously, drinking water has long been part of the practices enabling human existence—some would say an essential part. But the ways in which this practice has been rationalized, carried out, and understood are subject to change, as are techniques for making up personhood. Following the emergence of principles of hydration, people have come to drink more water, and to drink it differently—or so we venture to state here. But how has this happened? How have people become attached to bottled water? And how does this development depend on wider connections to techniques of self-practice and self-assessment that have gained currency since the late twentieth century?

Hydration has become a major theme in bottled water advertising. It allows companies to appeal to biomedical ideas of health in order to position their product as an essential part of self-health and healthy lifestyles. Alongside related principles such as the eight-times-eight rule (the idea that it is necessary to drink eight eight-ounce glasses of water a day), the focus on hydration has done much to establish new practices of drinking water, in which the consumer appears to be *always at risk of dehydration* and so must exercise constant low-level vigilance by preparing themselves for frequent sipping. Of course, scientific information is often featured on food and beverage packaging—in the form of nutrition labeling, for example— whence it enters into consumers' health calculations. But references to hydration are distinctive in that they are not a description of the contents but call on a set of principles that have a life independent of the brand. Like dieting and nutrition, hydration enacts a practice or "regime of living" (Collier and Lakoff 2005; Ong and Collier 2005) that is both framed by and instituted through biomedical vernacular. Moreover, promoting the principle of hydration appears to be tantamount to promoting the product itself. Creating attachments to this product entails identifying and fostering new habits among consumers, and specifically, frequent sipping to prevent dehydration. In this sense, hydration science is a critical part of the event of bottled water; it has been key to the formation of the relevant market attachments. Only on the basis of this principle does carrying a bottle of water around begin to seem like a logical and necessary thing to do.

The biomedical framing of hydration connects bottled water to certain ways of conceiving and acting on personhood that have become particularly significant in contemporary society. Novas and Rose (2005) have coined the term "biological citizenship" to refer to the way in which biomedical languages are increasingly used to understand and describe aspects of the self and others in late twentieth- and early twenty-first-century life. In this context, authorities increasingly conceive of individuals in biological and biomedical terms, which suggests the emergence of a new paradigm for making up citizens. But what is also significant is how "biologically colored languages" enter into people's descriptions of themselves, including their attempts to act on themselves, their predicaments, their aspirations, and their identity through processes that entail forging new relations with medical and scientific authorities and, one might add, health products. The overarching context for these developments is the recent and wide-ranging reorganization of the powers of the state, which resulted in many of the concerns that were once considered state responsibilities—the provision of welfare, health, and security—devolving to autonomous and semi-autonomous bodies (professional groups, the market, the family, the individual), which thus become responsible for managing their health, their welfare, and their affairs. Here, health is no longer cast as something that is imposed on subjects through governmental strategies and disciplinary techniques. Rather, it becomes a goal that must be actively embraced by autonomous subjects in the new domain of consumption, "ensured through a combination of the market, expertise, and a regulated autonomy" (Rose 1998, 162). Significantly, health is reconstituted in this process. It is no longer limited to the goal of preventing disease or prolonging life but incorporates various attempts to reshape, enhance, improve, and optimize the body—understood as a personal responsibility—which tends to require some investment in new technologies and commodities, and their promises. In this respect, the efforts of the beverage industry to promote hydration can be seen in the broader context of the emergence of technologies of personhood that have acquired wide currency within contemporary consumer society. But, while Novas and Rose point to the ways in which biomedical rationalities are taken up by entrepreneurs and industry in order to build markets, the connection between biological citizens and market elaboration has yet to be worked through in any detail. The story of hydration helps to reveal how biomedically inspired technologies of the self can be redeployed as market devices in the effective elaboration of markets.

The story of hydration is also a story about the diffusion and circulation of a "scientific" principle—indeed, of public engagement with science. How has this relatively technical health-related concept acquired such widespread currency and become such a familiar part of everyday languages and practices of the self? While beverage industry efforts to promote the concept are clearly important here, they are only one part of the story. To gain traction, this idea has had to connect more intimately and extensively with the activities and meanings of bodies. On this count, we argue that an inquiry into the making of scientific facts—not merely their consumption and communication—provides important clues about their salience or eventual capacity to engage publics. In particular, we need to situate the

producers and consumers of scientific principles and products as embodied—that is, as already caught up in meaningful practices, activities, and
relations—and look at their interconnection in processes of emergence,
rather than track the transmission of fully formed concepts between already
fixed entities (whether science, consumers, or industry). In this respect,
this chapter situates the field of health, fitness, and exercise science, from
which scientific principles of hydration first emerged as a dynamic zone of
popular culture in which bodies have "learned to be affected" by scientific
concepts, practices, and principles in the very process of their elaboration
(Latour 2004). This is not to debunk these scientific principles but rather to
situate them—that is, to approach their articulation as *events*, which do not
involve the application of human categories to the passive, natural world
so much as a modification of all the components or circumstances of the
event through a process of articulation, or what A. N. Whitehead (1978)
calls "concrescence." From this perspective, the emergence of new principles of hydration can be seen to rest on a particular articulation of bodies,
technologies of government, exercise science, and the beverage industry in
which what gets proposed is that bodies will perform better through practices of frequent sipping. The experience of thirst, conceptions of health
and responsibility, and practices of self-performance each become altered
in the process. Part of the event of bottled water, then, is the emergence
of new relations to drinking and new relations to thirst: bodies and their
vital exchanges begin to get organized and experienced differently. This
approach to the scientific event enables a more active grasp of the ways
in which contemporary subjects of hydration have been (and are being)
assembled. It enables us to develop a more extensive response to the question, how have we become so thirsty?

Hydration "On the Go"

To make sure you are well hydrated day and night, wherever you go and whatever
you're up to, evian® has a convenient PET bottle that's right for you. At work or at
home, opt for the larger formats. For sports, travel, or life on the go, the smaller sizes
fit easily into your purse, pocket, briefcase or knapsack.
—Evian (2014)

In previous chapters, we saw how PET plastic helps to equip the subject
of bottled water: the subject "on the go." This aspect is borne out in the

advertisement for Evian cited here: where the traditional glass bottle is proposed "for elegant occasions" and "to enhance fine dining experiences," the PET bottle—described as easy enough to fit in your "purse, pocket, briefcase or knapsack"—is said to befit "sports, travel or life on the go."[1] Different materials intertwine with different meanings and occasions, allowing different modes of consumption to emerge. Spearheading this appeal to convenience and mobility, a certain discourse is conspicuous here. The idea of being "well hydrated day and night" gives this mode of consumption its practical logic, its subjective rationale and form. The concept of hydration anticipates and suggests a particular manner of relating to water, enacted through specific practices, materials, movements, and techniques. Here we can see how ideas of health combine with certain materials, such as PET plastic, to produce a distinctive cluster of meaning and practice: a specific rationale for consuming more water. But where does this discourse of hydration come from? What are its associations? And what does it do for bottled water?

"Hydration" is notable for the way it conflates several disparate ideas about the relation of water and health. Initially emerging out of exercise science, where the dominant concern was the effects of fluid loss on athletic performance, the concept is now used much more promiscuously by the bottled water industry to evoke several quite distinct senses of health and the value of drinking water. These in turn have quite different medical genealogies. Alongside the discourse of replacing fluids that pertains primarily to the active and aerobic body, it is now common to encounter elements of a discourse of detoxification (concerned with ridding the body of toxins such as alcohol, tea, and coffee); a discourse of weight loss and diabetes prevention (where water is positioned as the healthy alternative to soft drinks); a discourse of renal health, concerned with the health and functioning of the kidneys; and a discourse of heat stress, concerned with the effects of heat on the body. An appeal to dermatology involving the promise of clear skin is also apparent in some instances. Hydration is in this sense a truly heteroglossic concept, in the sense that it embodies diverse medical voices—and therein lies its power: its ability to mobilize and attach itself to any number of (not necessarily commensurable) concerns and quasi-scientific ideas about the body's optimal health, and render the consumption of the product a means of meeting the body's requirements.

But it is also useful to distinguish hydration from some of the other health discourses that have animated markets in bottled water. Hydration is different from the discourse of therapeutic minerals, for example, which values bottled water for the special properties of the particular compounds it contains. This discourse, which has traditionally informed the use of table waters such as Perrier and Evian, draws on a longer history of hydrotherapy and "taking the waters" that emphasizes the distinctive properties of minerals from particular springs and spas (see van Tubergen and van der Linden 2002; see also chapter 2).[2] Although this tradition was an important precursor to the mass markets in bottled water that developed in the 1990s, it appeals to quite different therapeutic logics and principles than the discourse of hydration. Bottled water products are indebted to this tradition, but hydration adapts and transforms it, such that water's very status as generic—as a basic necessity—becomes the attribute that is said to give rise to its therapeutic qualities.

Hydration is also different from the discourse of risk, which bolsters the demand for bottled water from other directions. Where the discourse of risk is linked to anxieties around water contamination—and typically assuaged by appeals to purity in the branding literature—hydration is less concerned with the source of water supply than with the condition of the subject. It evokes a much more active or positive sense of health *promotion:* something that you do for yourself, akin to taking minerals or vitamins. But unlike the discourse of mineral water, this goodness does not consist in any special qualities of the contents themselves (beyond their generic identity as water) but rather in using the product on a regular basis, as an actively adopted practice. Thanks to hydration, we are no longer brought to buy water on the basis of the special properties of the minerals or elements it exclusively contains but because buying water is an especially prudent and healthy thing to do. To ensure proper hydration, moreover, it is necessary to have water constantly on hand. In this sense, hydration has been central to creating consumer attachments not only to water but to a specific way of carrying it that makes it instantly available. The concept of hydration has eased the passage of bottled water from a boutique to a mass product, all the while retaining its special (and especially active) association with health. The contents of the bottle may not be particularly special or distinctive, but the *practice* of carrying a bottle around and taking frequent sips from it promises to make *you* so.

Hydration can thus be located at the junction where discourses of biomedicine and lifestyle-oriented consumption converge. Here it is not simply the avoidance of sickness but the optimization of health—its continual enhancement—that constitutes the broad focus of health practice. Joseph Dumit (2002) has coined the term "dependent normality" to refer to an emerging paradigm of health, illness, treatment, and normalcy, in which it has become common to believe that our normal state is one of deficit and that healthy functioning can only be achieved by the use of remedial products. Dumit is mainly concerned with contemporary pharmaceutical reasoning—a far cry, it would seem, from a product as benign and basic as water. But a quick look at the construction of hydration in health advertisement information reveals several analogous features.

Dumit argues that a paradigm of "inherent illness" has begun to replace the paradigm of "inherent health" that once featured as the most common way of talking about health and illness. Where previously illness was seen as a temporary interruption of a default state of health, today the template of chronic illness has been extended to a wide range of conditions, such that "the normal state is one of vulnerability and precariousness, requiring a constant vigilance for further warning signs" (Dumit 2002, 125). In medical discourse, these warning signs are typically charted using algorithmic graphs that depict various symptoms as risk factors for the syndrome. In this model, one is always progressing toward the disorder, and the chart functions to dramatize the risk of discontinuing the remedial plan. Such a template of chronic illness infuses life with a pharmaceutical logic, in which drugs become the natural or obvious solution to a range of health problems conceived as imminent or chronic.

The discourse of hydration as a health amplifier depends on similar graphic mechanisms and calculative procedures. Although this anxious iconography would seem to be a long way from the notion of lifestyle enhancement or optimization mentioned earlier, the construction of dependent normality has been key to the naturalization of remedial products as essential components for extending and improving life. The circulation of these features in advertising, online media, and popular discourse has proved foundational to the making of new relations to water (see figure 3.1). As the appearance of this promotional information in a private health insurance magazine suggests, the discourse of health optimization resonates with notions of lifestyle, consumption, enhancement, and privatized provision.[3] Through these mechanisms, drinking water is reconstituted from a public resource into a private good or personal possession that promises not only to keep you well but to make you better—a personal responsibility.[4]

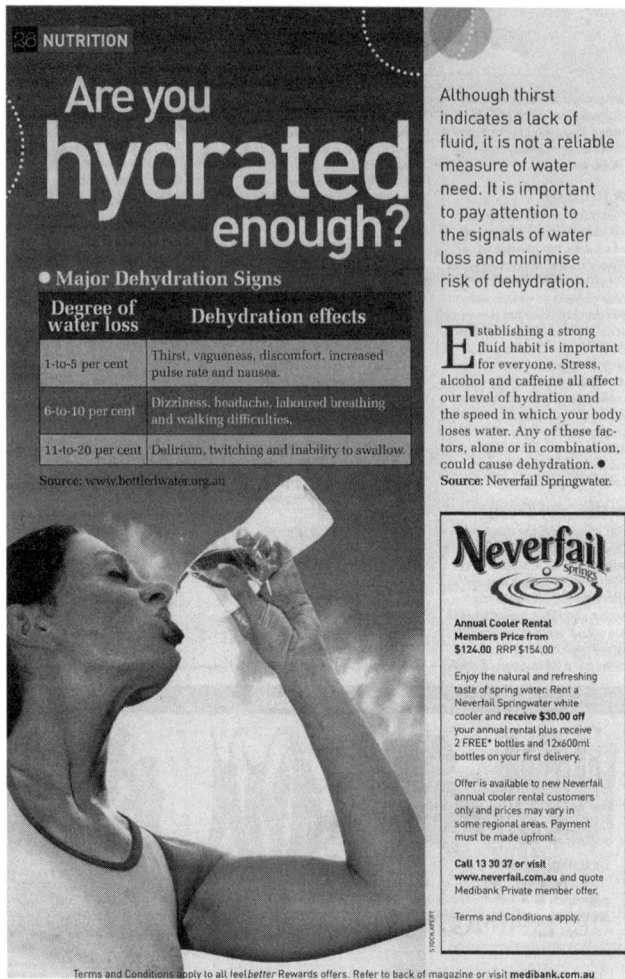

Figure 3.1
An advertisement in *Medibank Australia Magazine*, a healthy living magazine, 2009.
Source: Coca Cola South Pacific.

Aerobic Bodies

The transformation in water's ontology thus reorients our relation to concepts and regimes of health. The goal of "meeting the body's needs" is supplemented by a new emphasis on enhancement and performance. In this respect, the disciplinary context of the concept's history is especially revealing. For hydration science can be traced to the findings of the nascent discipline of exercise science rather than mainstream medicine, as might be expected. Prior to this moment, no one had found the issue of hydration that interesting except a team of research scientists who, on learning that U.S. troops might be required to fight in the desert, were driven into the Colorado desert in the 1940s to study the physiological effects of water depletion during physical activity (Adolph 1947).[5]

It was not until 1970 that anyone paid the topic much more attention. At that time a number of factors converged to make the topic newly interesting. The professionalization of sports prompted researchers to begin to apply scientific methods to physical exercise, and in particular to questions of athletic performance. David Costill, a researcher from Ball State University, is renowned for being one of the first researchers to apply scientific methods to the study of exercise and training in this context (Kolata 2001). At the time, marathon rules forbade runners to take fluids before they had run ten kilometers, and runners trained themselves to ignore their thirst while training (Cantwell 1980). But athletes were collapsing with body temperatures of up to 43°C and losing up to five kilograms on a single run. There was little money available to study sports at the time, so Costill approached the newly established beverage company Gatorade to ask for money to study the effects of dehydration on marathon runners. Not only did Gatorade agree, it provided enough funds for Costill to set up his Human Performance Laboratory at Ball State University, which still functions as a research and training hub today.

In his experiments on hydration, Costill can be understood as articulating certain propositions about the relation between the athletic body and its intake of fluids. As Latour (1999, 2004) explains, a proposition is not simply a claim or a statement; it involves articulating a body into a new set of arrangements or relations. Costill's experiments found that taking fluids (not only Gatorade) allowed runners to run farther and faster. By documenting that distance runners were in dire need of fluids, Costill's studies had the effect of prompting a major revision in marathon rules, as well as changes in training practice. In other words, Costill was engaged not so much in the discovery of some essential or universal truth about the body as in a quarrel or wrangle with existing habits of professional sports.[6] During the course of his initial experiments, Costill found that it was "not unusual for the extremely dehydrated runner to experience very little or no desire for water. The athlete must be aware of his body's demand for sodium chloride, water and glucose and realize that the thirst mechanism is an inadequate indicator of bodily needs" (Costill 1968, 35). On this basis, he went on to propose certain practical recommendations: "Because of the difficulty experienced by the runners while ingesting fluids and the discomfort caused by large amounts of fluid in the stomach, small frequent feedings appear to be most efficient and effective" (35). Despite the particularity of these recommendations to the circumstances faced by distance runners, in the hands of the beverage industry, the displacement of thirst and the recommendation of small frequent feedings went on to become standard features of hydration discourse generally ("frequent sipping").

These were the precursors of hydration science. But how did these principles travel and achieve the currency they have today? Costill's experiments coincided with an explosion of interest in running and aerobic activity in North America. Activities such as running specifically, and health and fitness pursuits more generally, grew from an elite practice to widespread and immensely popular interests and pastimes. The televised win of the Olympic marathon by American long-distance runner Frank Shorter in 1972 inspired many Americans to take up running and jogging. Indeed, some commentators attribute Shorter's win to the new protocols of training and hydration instituted by Costill's experiments—an interesting moment of synergy in the emerging assemblage of popular health and fitness. Public marathons such as the New York City Marathon (held annually since 1970), the Chicago Marathon (since 1977), and the London Marathon (since 1980) created public venues for the practice that were at once spectacular, eventful, participatory, and increasingly popular. People's enthusiasm for running was catered to by magazines such as *Runner's World*, which went to monthly publication in 1973, and books such as Kenneth Cooper's *Aerobics* (1968) and James Fixx's *The Complete Book of Running* (1977), which provided extensive practical information and advice for those interested in taking up these forms of exercise. Over this period, millions of people took up running, and this convergence of media, popular practice, and exercise science created new modes of engagement between consumers and scientific discourse around questions of bodily performance.

The principles of fluid replacement were a case in point. They demonstrate how the discourses and practices of exercise provided new ways for fitness enthusiasts to "learn to be affected" by scientific experiment (Latour 2004). In his bestselling *Complete Book of Running*, for example, Fixx enthuses, almost lyrically:

Most of us don't think a great deal about how much we drink or when we drink it. Runners however ... know that physical efficiency drops substantially if they drink substantially less than they lose. Instead of drinking merely because they feel thirsty (or not drinking because they don't feel thirsty), they drink consciously, deliberately, for the good of their bodies. They have learned how much their bodies need, and how often. What was once a mindless indulgence is elevated to an art. (Fixx 1977, 31)

He goes on to provide the relevant practical instruction. Here we can see the formation of a new cluster of meaning, practice, and identity. Fixx's book did a great deal to popularize the practice of running to millions of Americans. But what is especially significant is the articulation of the body with certain prehensions from scientific discourse. In this moment, the question of fluid intake is disconnected from the body's experience and rendered a matter of expert calculation and determination. Drinking is reconfigured in the process. No longer a simple response to how the body feels, it takes the shape of a health practice or skill: an activity that is undertaken "consciously, deliberately, and for the good of the body" (Fixx 1977, 31). Moreover, it can be learned and studied and optimized to enhance the body's efficiency and performance. Thirst is not what it used to be.

There are several features worth noting about this event that help situate the popular engagement with hydration. First, these principles emerged from exercise science, and not medical science or nutrition, as might be expected. This is the study of the body in motion, the body's performance, in quite specific—indeed, extreme—conditions. It does not pertain to the sedentary body—about which no reputable evidence justifying the eight-times-eight rule can be found, incidentally (McCartney 2011; Valtin 2002). Yet the disconnection of fluid needs from the experience of thirst went on to become a universal feature of the popular discourse, thanks to the beverage industry's marketing activities.

Second, the articulation of these principles came at a time when there was unprecedented interest in and uptake of activities that had previously been the domain of a few.[7] The groundswell of interest in running and aerobic exercise propelled unprecedented popular engagement with new concepts of the body, including scientific explanations of performance and especially differences in performance. The popular media that sprang up to cater to this new movement eagerly reported and discussed these findings. In other words, this was not a case of unidirectional science communication but something much more multilinear and coemergent. Costill

was renowned for asking the same questions as athletes and people trying to get fit: the scientific interest in the aerobic body was a complement to widespread popular interest. In this sense, the exercise press can
be understood as a key venue for the articulation of new ideas about sports
performance, informed by an ethos that was at once experimental, popular, scientific, participatory, and practical. These convergences allowed the
principles to travel.

Third, health and fitness had begun to serve as a compelling source of
identification and symbolism much more broadly at this time. This development may help explain the interest in performance and hydration beyond
the specialized community of professional athletes. Robert Crawford has
discussed how health and fitness became a key vehicle in this period for the
expression of certain values with which the professional middle classes had
long identified: personal responsibility, self-discipline, self-determination,
and willpower (Crawford 2006). As a form of work on the self, running and
other forms of aerobic activity seemed to provide a visceral model of what
hard work and discipline would differentially bestow. While competitive
opportunities had always existed in sports, running fostered a new kind of
competitiveness, in which individuals could compete against themselves.
The notion of the "personal best"—the idea that you could draw out your
own potential through competing with yourself—provided a practical frame
that could be used to make sense of the competitive nature of white-collar
work, especially during the economic downturn of the mid-1970s. If the
domain of the personal and the corporeal had been a primary beneficiary
of the liberation movements of the postwar period, in conditions of world
recession it was increasingly constructed as precisely the object on which
one had to work to maintain security and control (Race 2009). For many in
the professional middle classes, exercise came to feature as a key medium
through which the values that would later underwrite neoliberal philoso-
phies of the self were both expressed and embodied, as a sort of "practical
sense" or habitus (Bourdieu 1984). For others, an aspirational identification
with this figure would suffice, as the active self expressed through the aerobic body came to substitute for the ideal of normative citizenship itself. The
aerobic body came both to analogize and displace the overworked body—
the stressed body, the body under duress—in an equation that bottled water
advertising would go on to make much of in the 1980s and 1990s (witness,
for example, the appeal to "life simplified" in Dasani marketing, discussed

in chapter 2: the sense that water might alleviate the pressures and stresses faced by the multitasking body and replenish it). The coolness of water—a quality especially afforded by the refrigeration of bottles—is, of course, also significant here, for if the body and its stresses are understood and experienced in thermodynamic terms, coolness becomes the affective equivalent to refreshment and replenishment.

In short, the popular identification with health and fitness that emerged at this time might be regarded as "an attachment to a cluster of promises" (Berlant 2011, 23) that allowed people to imagine belonging to something or having a future, or simply to organize their lives by providing a frame for deriving a sense of accomplishment in relation to small, personal goals.[8] Whether this identification bore any strict or direct correspondence with the actual fitness activities of all those who came to invest in it is less significant and also less certain. It is the aspirational attachment to health and fitness as a source of identity and belonging that contemporary bottled water advertisements rely on when they depict the sporting body as the absolutely generic body of the consumer public: a body that is at once normative and aspirational; presumed as general yet slightly out of reach; what everyone is supposed to be in the regular, everyday, normative mode of being "yourself, only better" (to take from the Fitness First slogan) that has come to characterize healthy consumer citizenship today. The tendency of the bottled water industry to extrapolate and generalize a particular body to the body of the general public is evidenced, for example, in the self-assessment tools found on the International Bottled Water Association (2014) website that urge consumers to calculate "how much water your body needs." The website features a hydration calculator where the visitor is asked to enter weight, exercise minutes, and exercise intensity. Even if the visitor enters a value of zero for the two exercise questions, she or he is taken to a page featuring advice from the National Association of Athletic Trainers, which states, "When exercising vigorously, you need to begin in a well hydrated state." This prescriptive identification does much more than tell us about the identity of the idealized or presumed consumer. As we shall see, calculation tools such as these also connect with techniques of the self that have a much broader purchase in contemporary society, tapping into self-assessment techniques that have become a common feature of people's everyday engagements with biomedicine and practices of "healthy citizenship" more generally.

Biocalculating Bodies

For Joseph Dumit (2002), the biomedical habit of visualizing and chart-
ing risk sets in motion a particular logic of illness that naturalizes the use
of therapeutic products. What is conspicuous about the contemporary
context is the way these mechanisms for picturing and calculating illness
and risk are actively embraced by consumer citizens to make sense of their
conditions. The Internet has proved a particularly significant mechanism
for facilitating these practices of patient self-education and biomedical
engagement. People use online resources to develop quite specialized scientific and medical understandings of their being—a practice Novas and
Rose (2005) term "informational citizenship." And they form themselves
into communities linked electronically by email lists and websites around
particular conditions—a development Novas and Rose call "digital biocitizenship." For Novas and Rose, these forms of engagement with biological
explanations of the body are a distinctive development in the process of
caring for, or caring about, health: "An active citizenship is increasingly
enacted, in which individuals are taking a dynamic role in enhancing their
own scientific—and especially biomedical—literacy" (446).

Biotech, biomedical, and pharmaceutical companies have been particularly quick to tap into the self-education practices of biological citizens.
Websites designed by pharmaceutical companies are geared toward the
practices of active health consumers, providing educational material, scientific explanations, self-assessment tools, and practical tips and advice,
installing ways for individuals to understand their condition and imagine
how pharmaceutical agents will target it. Legal changes that allow pharmaceutical companies to engage in direct-to-consumer advertising in the
United States provide the general context for these marketing initiatives.
But in jurisdictions where direct-to-consumer advertising is not permitted,
companies still employ the Internet for marketing purposes. Now, however,
the information is presented not as an advertisement for a particular product but as a resource for self-education about the condition itself. There
are now both a marketing literature and a critical literature on "condition
branding" (for example, Moynihan and Henry 2006), in which "trust in
brands appears capable of supplanting trust in neutral scientific expertise"
and "biovalue ... is supplanting public value in the biological education of
citizen-consumers" (Novas and Rose 2005, 448; Waldby 2000, 33).[9]

In this respect, the self-assessment tools found in this domain can be understood as market devices, much like the bottles and brands discussed in previous chapters. As market devices, they are seized on by industry to make certain transactions more likely or probable. By offering apparently authoritative procedures for assessing the self and establishing need, they replace the fickleness of consumer desire with a specifically formatted setup replete with calculative tools and designated procedures. This makes them invaluable in the construction of markets. The ability to calculate is not concentrated on any individual actor here but, crucially, is distributed among several actors, devices, and material settings, such that the shaping of markets inevitably involves the adjustment and articulation of diverse sites of agency and calculation. This is *agencement* at work (Callon, Millo, and Muniesa 2007, 2). In observing the link between techniques of biomedical self-assessment and market devices, we are a long way from the human-centric theory of "false needs" imagined by the Frankfurt School in its theorization of consumption, with its focus on ideological consciousness, inflated objects, and the inciting of false desires. We are also a long way from conceptions of subjects entering the market with preformed identities and making consumption choices expressive of these identities. As we have argued, the shaping of markets involves the continual massaging and modification of diverse sites of agency—objects, humans, and material procedures—each of which is already implicated in practices of formation that are already "on the go," as it were. These efforts may involve putting in place, promoting, or modifying particular calculative practices, for example, that, while not determinative, lend themselves to particular market choices and aim to produce attachments.

The Hydration for Health Initiative (H4H; 2014) website provides a particularly vivid example of these calculative devices and their connection to techniques of self-assessment that are already implicated in the making of biomedical subjects. Indeed, the website puts itself forward as an eager and obliging participant in the self-assessment practices and thus self-formation of subjects. Sponsored by Danone, the website is divided into two sections, one for "members of the general public" and one for "healthcare professionals." The first of these draws heavily on the above-mentioned genre of medical education. It features a high-definition hydration calculator designed to "help you see whether you're getting enough fluid in your daily diet." After entering details of one's weight, height, age, lifestyle, gender, and country, the user is asked to enter detailed information about daily beverage intake—an item that is further discretized to no less than ten different categories of beverage, from soda to sports drinks. The results page shows a calculation of whether the user has good or bad hydration, and conveys warnings about the fluid losses associated with different activities, including advice on what to do to compensate for them. A sidebar provides information on the intake of liquid calories from different beverages, stating that "water is the only liquid necessary for proper hydration."[10] By means of this device, the need for water is rendered calculable for consumers in a manner that bypasses traditional mechanisms used to register the need for water, such as thirst. Consumers are invited to engage in particular techniques of self-assessment, enacting a form of calculative agency steeped in biomedical authority, the likely outcomes of which are factored into producers' and marketers' equations. In other words, consumers are led to engage in processes of biocalculation that are specifically proposed to create attachments to the relevant goods and processes.

But wait, there's more! A third section of the public website features a high-gloss informational animation titled "Water and Your Body." By choosing a profile (the choices are Over-Indulgent, Pregnant, Senior, Active, or Children), the user is presented with narrated scientific and medical information especially tailored to her or his personal profile.[11] Using scientific terminology, the narrator explains in extensive detail why hydration is a necessary concern for the visitor. For each profile type, detailed information is provided on how different fluids and activities affect different organs of the body—the pancreas, kidneys, fat cells, heart, muscles, lungs, and brain. These "health need" profiles are used to constitute different market segments and match them with the different medical concerns that can apparently be resolved by drinking water, making further use of the heterogeneity of hydration discourse. For example, the Over-Indulgent category contrasts water with sugar-sweetened beverages, while the Active category features information on fluid loss during physical activity and thermoregulation—another instance of heteroglossic opportunism.

Meanwhile, the section of the website for health care professionals boasts an extensive range of features, from the mission statement ("to establish healthy hydration as an integral part of public health nutritional guidelines and routine patient counseling so people can make informed choices"), to the delivery of scientific information about the health value of water, to the

Hydration for Health Hub, "an interactive portal to the healthy hydration community worldwide ... intended only for health professionals." Remarkably, the site also offers a range of tools and publications for use by health professionals, including downloadable educational slide kits (designed for use in professional and educational conferences), hydration charts, patient education materials, and online patient advice; and a list of congresses and events, which include the annual scientific H4H meeting, sponsored by Danone and held every year in Évian since 2009.[12] Long gone are the days of marketing initiatives targeting consumer desire exclusively. Here we have the corporation's participation in, and attempted shaping of, a whole field of knowledge and practice. Separately, the payoffs of each specific intervention in this field may not be direct or guaranteed, but together they form part of a complex assemblage of discourse and practice that creates connections between scientific processes and the elaboration of market selves with specific dispositions and attachments.

Critics of commercial influence have documented the emergence of informal alliances among pharmaceutical corporations, public relations companies, journalists, doctors' groups, and patient advocates in the contemporary biomedical field. From the perspective of pharmaceutical corporations, these alliances are effective not only in publicizing little-known conditions but also in fostering the creation of entirely new medical disorders and dysfunctions, and promoting these ideas to policymakers and the public (see, for example, Moynihan 2003). One can see the adoption of many of these practices by Danone in its sponsoring of scientific meetings, enrolment of experts, and adoption of the language of health policy activism in its mission to influence public health guidelines. But this is not a straightforward process of product promotion, for it entails seizing and negotiating moments of opportunity in an unfolding assemblage of biomedical, economic, and bodily production. Also noteworthy here is the complete absence of any images of branded bottles on either website, including any mention of products such as Evian at all. All the images of drinking water on the website feature glasses or unmarked bottles. Each page carries a visible claim, "Sponsored by Danone Waters," but this is all that the company will venture, to maintain legitimacy under the terms and conventions of health education discourse. To recall Novas and Rose (2005), the corporation is engaged here in a complex process to have the brand associated with biomedical principles and scientific legitimacy in a field where critical scrutiny of commercial influence is increasingly promi-nent (see Moynihan and Henry 2006). The corporation does not determine or control this process but anticipates and suggests connections between the self-constituting practices of subjects, authorities, and technologies that are already under way.

What we have here more generally is a stunning convergence of bever-age advertising with the genres and practices of pharmaceutical marketing, which sees beverage producers taking up many of the practices pioneered by pharmaceutical corporations, not only to position their product but also to participate in the making of biomedical knowledge. It would be a mis-take, though, to write off Danone's efforts as bogus science, or distinguish them from legitimate science merely on the basis of commercial influence. As Moynihan and Henry's (2006) work demonstrates, corporate sponsor-ship of R&D is a regular feature of the contemporary biomedical field. As we have attempted to demonstrate, what is needed is ongoing attention to the historical and material coconstruction of markets and biomedical facts, with attention to how calculative procedures are constructed and enacted by the bodies that participate in and emerge from these encounters.

Fluid Habits

Brawndo's got what plants crave. It's got electrolytes.
—*Idiocracy* (Judge 2006)

The film *Idiocracy* depicts a world in which hapless citizens kill their crops by irrigating them with a commercial sports drink because "it's got elec-trolytes." The film caricatures a slavish relation to advertising slogans and bioscientific language that some might be tempted to recognize in today's consumer practices and cultures. But the film's dystopian vision of the future is predicated on an idea of consumer stupidity that seriously under-estimates the distributed character and hybridity of market devices. As this chapter has shown, these devices are performative: they work by prompting and suggesting, rather than simply manipulating; by making connections; by tapping into much more widely invested techniques of making sense of, and acting on, persons. An understanding of markets, facts, and subjects as mutually constitutive, distributed, and emergent produces a more sat-isfactory relation to technoscientific production—and a more satisfactory

account of active citizenship—than a commitment to consumer stupidity would allow. To grasp what is happening here, we need more than an
analysis of meaning and representation as advocated by cultural studies of
consumption, and more than an analysis of commodity chains as advocated by the anthropology and sociology of consumption. We need to be
able to understand the multiple forms of technical labor that go into the
production of consumer attachments; how the product becomes associated, both implicitly and explicitly, with already existing technologies of
self, such as patient self-assessment tools and self-diagnosing practices; and
"regimes of living" (Collier and Lakoff 2005), such as biological citizenship
(Novas and Rose 2005). And we need to understand how technical labor
forms connections with other meaningful practices, and with techniques
for making sense of those practices (such as exercise science), that have
come to organize and invest the body and its performance in wide-ranging
ways. Markets seek out dispositional tendencies and techniques of the self
that already have purchase and already are in process, latching on to these,
shaping and fostering them to create attachments between goods and persons. In this respect, the techniques of the self that characterize biological citizenship do not simply feature as convenient opportunities for the
promotion of products. Bottled water, personal health calculators, online
self-assessment tools, and branded conditions have become common components in a whole range of contemporary practices of self-formation. Such
devices function as orienting devices, promoting certain habits in the making of subjects.

Conclusion

In standard accounts, the emergence of bottled water as a commodity is
further evidence of the inexorable logic of capital—another chapter in the
relentless commodification of goods, people, and resources that characterizes contemporary economic regimes. And, from a certain perspective, it is.
However, in this chapter we have chosen to adopt a different approach—
one that treats market objects as unfinished and entangled, that pays attention to the situated ways in which markets are assembled in and through
practical interaction. We believe this attention to *how things are arranged*
and to the contingency of events affords access to multiple other insights
and stances. It helps generate a precise but more open-ended account of

how market *agencements* come to have effects in the world. This approach helps to grasp these effects in their complexity, in ways that are grounded in historical and practical situations, and this has the advantage of configuring markets as malleable, complex, diverse, and evolving. Nothing is predictable about how markets develop; they are subject to a diverse range of practical operations. They encounter obstacles and challenges distributed across vast and evolving networks that are addressed through concrete and practical trials and negotiations. What eventuates from such processes is not totally determined or inevitable but the result of how these encounters are resolved and how they unfold in their particular specificity.

In this opening section of the book, we have explored three events or historical developments that served as key devices in the assembly of mass markets in bottled water. Without them, bottled water would simply not be the object it is today. The development of PET plastic was a key moment in this process, useful for packaging water and providing new ways of literally and metaphorically grasping it. As the outcome of complex informational relays between industrial objectives and the material performance of plastic, the invention of PET was far from inevitable: it was a contingent event that made a range of new things possible. Most obviously, it enabled water to launch itself into retail arrangements as an intelligible and calculable item, and it created opportunities for water to enter into new forms of circulation. But PET also conferred new properties on water, accentuating certain qualities and making them marketable and available to commercial articulation. As a hybrid object, bottled water is personal, disposable, contained, discrete, and easily transportable, and these attributes reflect on other sorts of water arrangements. The encounter between PET plastic and water also provoked other forms of action, apart from market ones: it generated political affects, dramatizing wider tensions between the state and corporate power in the distribution of biopolitical resources. As we discuss in later chapters, political agencies seized on this conjunction of plastic and water to launch protests that continue to play a major part in shaping the forms these markets take today.

Branding is the second element we identified as a constitutive event for this commodity. Without branding, bottled water would be an ineffectual market thing. But, in our account, branding is more than a matter of glossy semiotics. The brand is a complex ontological object that shapes relations between producers, authorities, consumers, calculative agencies, and more.

More than simply a sign of corporate agency, the brand relies on the regulatory practices of state authorities, among other actors, to enjoin trust,
confer value, and create attachments, as the story of Evian suggests. As a
public object, the brand is never stable but must be continually managed;
it is subject to diverse trials and procedures of qualification. The branding
of water also reveals how brands perform as political devices: in qualifying
bottled water: they qualify or disqualify (by omission) other forms of water
provision. In these maneuvers, unbranded water emerges as devoid of precisely the qualities claimed for bottled water. Branding thus reconfigures
biopolitical processes by inserting new economic relations and procedures
into their arrangement.

The account of hydration science and its emergence developed in chapter 3 shows how scientific production generates new forms of personhood
by promoting particular techniques of self-formation. Increasingly, markets
are not simply reliant on these techniques but feature as key constituents
in their practical operation. Markets work by creating attachments that
insinuate themselves into practices of making up persons, whether given
or emergent. But it would be wrong to imagine that hydration science is
simply the cynical invention of the beverage industry. It emerged from a
complex constellation of actors, habits, and processes that were brought
together in experiments designed to optimize human performance in professional sports. The diffusion of hydration science to a much wider public
reveals how opportunistically the beverage industry fosters market attachments in the context of already popular practices such as self-health. Tools
of self-diagnosis are not only prominent in the practices that characterize
consumerized medicine; they offer quantifiable mechanisms and routinized formulas for creating attachments to health commodities. This makes
them appealing as market devices. Through these means, the subject of
hydration has become attached not only to water but to specific ways of
carrying it and to other habits—specifically, frequent sipping.

In historical terms, it is difficult not to be impressed by the seeming inge-
nuity of these developments, by the sheer diversity of factors, elements,
and designs that might be taken historically to have conspired to produce
this market object. But this is not a teleological story, and we want to refute
the impression that what we are left with at the end of these processes is
a complete, stable, and settled object. For a start, these processes are not
over. One of the values of the event-based analysis we have pursued over

the last few chapters is the light it sheds on outcomes and effects that are both unexpected and surprisingly generative. The containerization of water has prompted new perspectives on health, new political allegiances, and new and unexpected associations between heterogeneous elements, well beyond the market frame. Various shadow realities have emerged alongside bottled water markets, for as a commodity, it equips persons in highly variable ways. If any doubt remains, in the next part of the book it becomes clear that bottled water is nothing like a complete or finalized object. It is enacted in practice, and these practices open it up to diverse realities and impacts that are complex, multiple, and ongoing.

II Bottle Practices

4   Drinking Water Arrangements in Bangkok: Accommodating Bottles

The rapid and relatively recent growth of bottled water markets was a product of diverse historical processes. Plastics R&D, brands, sports science, water quality concerns, gyms, drinkers, beverage corporations, and much more all played a role in assembling bottled water as an event (DeLanda 2006, 3). The extensiveness of this event, its actualization in specific local configurations, is the subject of this part of the book. Our primary aim is to understand how bottles of water have been actualized as *ordinary:* how bottles function practically, how they acquire meaning and value, and how they become normalized as a routine feature of everyday drinking and disposal practices. While it may seem that events are the singular or startling antithesis to the repetitive rhythms of daily life, this opposition is philosophically inaccurate. If event thinking pertains to becoming and effectivity, rather than structure or essence, as we argued in part I, then the "becoming normal' of bottled water is significant and eventful. For in this process new realities emerge that rearrange the relations between and among things, bodies, water, habits, and economies. The question then becomes, in what ways do these realities make the bottle appear as an effective solution to wider biopolitical concerns about water and the support of life?

In seeking to understand how bottled water becomes ordinary and the implications of this for biopolitics, we turn to examples from three Asian cities: Bangkok, Chennai, and Hanoi. This choice is deliberate. Not only does it allow us to investigate bottled water in a region of the world where markets are growing fastest (Drake 2010; International Euromonitor 2010; Rani et al. 2012; Rodwan 2011), it also allows us to explore bottled water practices in places where the water infrastructure is often underdeveloped, unsafe, or both—places where turning on the tap and encountering a regular, safe flow is far from being a universal experience. This unevenness in water supply and access plays out in vastly different ways in each of these cities. Our intention is not to discuss "Asia" in generic or totalizing terms.[1] Rather, we wish to acknowledge that the modernist governmental ideal of extensive reticulated water networks has never been realized in many parts of the region. Each of the cities we investigate has a distinct water history involving complicated and diverse associations between municipal and informal providers, complex sociotechnical arrangements, and various market regimes. In different ways, they reveal the dynamics of "splinter-ing urbanism," or what Bakker (2010, 23) describes as "fragmented urban water supply networks." These messy and also highly situated water reali-ties have important implications for bottled water markets and the ways in which water in bottles is both qualified and consumed, as well as its mate-rial impacts on worlds.

This urban water context thus reframes the bottle from a convenience or leisure consumption item into a significant participant in potable water provision. Of course, bottles are still drunk from on the go, but they are also increasingly incorporated into households and the daily rhythms of urban survival. They are part of the widespread growth of water hierarchies, or the use of different sources and qualities of water for different everyday functions. How, then, to explain the ways in which bottled water becomes normalized in these three cities? Bakker (2010) offers important insights into how to approach this critical question. She shows how inadequate many existing analytical categories are for understanding the complexities of urban water systems in developing and middle-income countries. Oppo-sitions like "private" and "public" or descriptors like "informal" are incapa-ble of capturing the mess and fragmentation of water supplies or the ways in which multiple modes of water provision coexist and interact (Bakker 2010, 22). Equally problematic is the critique of water markets—bottled and otherwise—as uniform evidence of "privatization" or neoliberalism. These assessments reduce the rise of the bottle to the logics of political-economic processes and make it difficult to understand exactly how markets in bot-tled water are assembled and the ways in which the bottle is made mean-ingful in relation to other forms of water provision and everyday practices. While bottled water markets are no doubt opportunistic responses to vari-ous forms of state failure and fragile hydrological environments, it is neces-sary to understand the *ways* in which these markets are enacted in relation to other water supplies, and the specific practices they prompt.

Rather than dismiss markets out of hand, the challenge is to make sense of the ways in which they become embedded in existing water cultures and come to interact with and shape these in specific ways. If our aim is to understand how bottled water markets become seen as a pragmatic solution to wider biopolitical concerns about access to safe water, then different conceptual tools are needed—tools able to tease out how various bottled water realities emerge and take hold. To this end, our main emphasis in this section is on the dynamics of practice, enactment, and ontology. These concepts, central to science and technology studies and process philosophy, open up the idea of market practices to a richer and more complex elaboration beyond economic exchange. They make it possible to understand how markets are more than simply economic in their impacts and certainly cannot be considered natural or immune from ongoing intervention and rearrangement. While the key functions of markets might be to organize exchange and establish calculative agencies between participants, these functions can be carried out in many different ways. Equally significant is the way in which consumers are constantly qualifying and requalifying goods and incorporating them into everyday lives, thereby extending and articulating the economic into dispersed and diverse cultures of use. Markets, then, do not constitute a singular reality that simply obliterates others; they are specific assemblages that create distinct "spaces of calculability" (Callon 1998, 191). In these spaces, agency is distributed and the differing calculative capacity of all participants emerges as an outcome of putting things into new valuing relations. These forms of value are never exclusively economic: they are multiple and shifting, and a result of diverse qualification procedures, as we have argued (see also Callon, Méadel and Rabeharisoa 2002). The question concerns exactly how markets enable water in bottles to be qualified as safer or more convenient than other forms of supply. Through what specific everyday and economic practices does bottled water become valued as a better way to access drinking water? These questions arise prominently in our first case study, on Bangkok. In the next chapter we investigate how the emergence of bottled water markets in Chennai, the sixth largest city in India, has become implicated not only in new drinking practices but also in the enactment of serious water scarcity. Finally, in chapter 6 we explore a shadow reality of bottled water market practices, the growing amount of plastic waste in Hanoi. In that discussion we look at the afterlife of the bottle and the ways in which market externalities generate their own calculative agency, their own capacity to enact new forms of value in the transition from waste to recycled plastic feedstock.

Ontological Anxiety about Water: Bangkok

As we saw in chapter 2, qualification procedures described the operations through which the value of spring water from Évian-les-Bains was enacted in relation to a range of metrological devices. These procedures and devices, from state regulations to branding, were central to the coming into being of this water as a market thing. In this chapter, qualification procedures refer to the diverse practices that consumers, producers, state authorities, and other actors use to establish the quality or safety of drinking water in Bangkok—practices that involve various degrees of engagement with bottled water markets. The relevance of Callon's (Callon, Méadel, and Rabeharisoa 2002) concept of qualification is that it draws attention to the specific techniques through which value and quality are enacted, and the ways in which qualification is always relational. In the case of Bangkok it is possible to trace how various ontological anxieties about the quality of drinking water developed and how the value of bottled water emerged in relation these anxieties. The key issue is that this qualification of bottled water relative to existing supplies was not simply imposed on gullible consumers by market forces or marketing hype. It came about *through use*, in and through complex daily practices whereby consumers and bottles and myriad other elements began to craft new realities and to produce new forms of ontological security and insecurity.

What makes Bangkok particularly apt as a location in which to explore this process is Thailand's status as a developing economy, with many features that have come to be taken as emblematic of the modern nation-state. Among these features are its social services and resource-management systems, including the state-owned water distribution system, which is overseen by the Metropolitan Waterworks Authority (MWA) in Bangkok. The MWA provides water services cheaply and efficiently to greater Bangkok, with a high level of coverage. In 1999, the quality of the state-owned water supply was independently assessed as complying with World Health Organization (WHO) standards throughout the MWA network. The MWA regularly monitors the quality of the water in its network and claims on this basis that water from the tap in Bangkok is absolutely safe for drinking. The municipal supply is also subject to ongoing improvement operations in its bid to meet international standards for water quality.

Members of the growing Bangkok middle class do not necessarily share this confidence in the quality of the municipal supply, however. Drinking water straight from the tap is generally regarded as a practice only of the poor or those who have no choice. Traditionally, tap water was boiled before being considered fit to drink, and in this respect water has long been subject to special treatment in Thai households. The relatively recent nature of government assurances about tap water has meant that special arrangements around drinking water tend to be the norm rather than the exception. As one of our informants whose family travels between Bangkok and Chiang Mai put it:

> We don't drink tap water in Bangkok—we use bottled water. Because taps not good enough for drinking but we can use it for other things. But, we are not sure. It is only just a few years ago that you can drink tap water in Bangkok. Not before that, you can't. (Small business owner, female, forty-two)

Despite state assurances about water quality and safety, the doubt and insecurity about the ontological status of tap water, which were typical among the middle-class residents we interviewed, create multiple openings for other devices, practices, and markets associated with drinking water to proliferate. Moreover, it becomes difficult to separate this ontological anxiety from other factors that bear on drinking water practices and purchasing habits, such as class aspirations, taste and distinction, performing modernity, personal commitments to maintaining health and well-being, choices in caring for others in the household, and convenience. The desires to avoid illness, demonstrate health, and improve one's social status are interdependent concerns that mutually constitute the "quality" of drinking water. Each of these concerns becomes available for the qualification procedures of markets, whereby producers seek to associate their goods with qualities whose value is relationally produced. Attachments to bottled water thus take shape within a dynamic social, material, and commercial landscape in which multiple drinking water objects, devices, claims, and practices vie for attention.

But there is more to say about the play of qualification and value as it bears on drinking water and devices in this context. Specifically, the process of arranging drinking water entails a certain amount of practical labor— whether inside or outside the household—and this labor may take different forms. In wealthy countries, the provision of quality drinking water from a tap is experienced as a form of convenience that residents of those countries take for granted. By "convenience," what is generally understood here is a reduction in the amount of time spent on ancillary chores—all those activities and tasks that would otherwise be associated with obtaining and preparing water so that it is fit to drink. This is one of the major affordances of the tap. As well as saving time, what is important in this concept of convenience, as Elizabeth Shove notes, is the ability to access the good (here, water) at very short notice: "In allowing users to 'store' time, defer activity or manage and minimize interruption, tools of this kind enhance capacity for autonomous organization" (Shove 2003, 172). In this sense, domestic technologies provide a greater degree of flexibility with regard to the events that make up everyday routines. Their relative value emerges in this context. To understand the quality of convenience, then, we need to understand how the relevant devices interact with, disrupt, displace, or simplify already established routines and practices of everyday life.

In this regard, what is needed is a better way of grasping the interactions of objects and practices, devices, and routines: to show how bottles, taps, bodies, and other devices become connected and interact with each other to create new ontological realities for drinking (Mol 2002, 6). For if values such as convenience are achieved relationally, it is necessary to understand the household *as it is practiced*—the practices, routines, and arrangements that characterize a given setting, which a given product or object proposes to simplify. Standard approaches to consumption are of limited help in this regard. These approaches pay insufficient attention to the materiality of the commodity (Bennett 2001; Brown 2003; Shove, Watson, and Hand 2007). They implicitly render the commodity a passive object of cultural inscription, a surface on which "culture" gets to work and makes meaning. In these frameworks, practices are largely things humans do to or with things to express identity, social positioning, or some deeper social order. Practices appear in these frameworks as relentlessly human and symbolic, emanating from human consciousness, intentionality, and discourse.

What is missing from these approaches is an understanding of how material things participate in the shaping of bodies, meanings, and practices—how the social is both produced and practiced in and through relations with artifacts, and is therefore not exterior to these relations. Hence the turn to science and technology studies (STS) in many accounts of practice.

For it is here that distributed forms of agency are recognized, and where the more-than-human hybrids of meaning and matter are central. Also important in this approach is the refusal to allow macrocategories such as "culture," "economy," and "society" any "trans-historical ontological validity," as Tony Bennett (2009, 102) says. In this analytical mode, reality is enacted or performed through the multiple relations by which things get associated. It is not a matter of identifying actions and practices as evidence of social forces or representations of deeper structures but rather of tracing how "the social" emerges in the dynamics of both durable and fleeting assemblages.

However, as Shove, Watson, and Hand (2007) argue, the STS approach still has limitations when it comes to developing a fully materialized account of practices and enactment: "The Latourian contention that artefacts literally construct socialness has yet to be worked through in any detail" (14). *The Design of Everyday Life* is their attempt to redress this with a close analysis of how various materials become implicated in everyday practices of consumption, from renovating to shopping. The empirical focus is on tracing the relations between and among objects, bodies, meanings, forms of competence, and routines in a variety of settings. What is especially valuable about this method is the commitment to understanding how material things and technologies become integrated into practices as performance, and how this both realizes their various material affordances and also, at the same time, stabilizes social relations over time. This does not mean that practices become fixed performances that are endlessly repeated. Practices continually evolve, integrating new elements and creating unexpected disruptions to existing routines.

One of the key distinctions Shove and colleagues (2007) make is between practice as performance and practice as entity. Practice as entity has a relatively enduring existence over time and space. It refers to the ways in which practices are made durable through the relationships between norms, materials, shared meanings, and bodily routines. Practice as performance is the specific enactment, the active doing through which practice as entity is sustained, reproduced, or changed. It refers to the contingent dimension of practices, the ways in which practices are both reproduced and continually reinvented in action (12–13). This theorization of practice pushes accounts of consumption beyond the registers of appropriation and domestication. These concepts implicitly endorse an approach in which humans tame things, in which the thing is incorporated into existing routines and spaces rather than being actively involved in making and remaking these. In other words, appropriation can imply a certain form of material stasis, or "socio-technical closure." Once the thing is stabilized or embedded in contexts, it remains relatively unchanged. The value of Shove and colleagues' approach lies in its insistence on the role of materials as actants that can suggest and transform practices—that is, on practices as complex assemblages of the human and nonhuman that are always on the move (8).

In our desire to make sense of the massive growth in bottled water consumption in Bangkok, Shove and colleagues' account of practices is extremely valuable. It directs attention to drinking as both practice as entity and practice as performance, and it emphasizes the role of the bottle as a commodity that participates in the emergence of new practices. Rather than critique the rise of the bottle as an environmental catastrophe or a threat to the provision of safe drinking water (the dominant tropes in many analyses of bottled water to date), a focus on drinking practices pays close attention to the ontological realities of bottles in action. The issue then becomes, how does drinking water *from bottles* emerge as a new practice? What is involved in this practice—what practical knowledge, routines, and norms sustain it? And what kinds of implications does bottled water practice as performance have for other drinking practices? How have bottles reinvented water, its technical delivery, and ways to drink?

These questions have a different political orientation from critique. They involve an *ontological* politics because they focus on how bottles come to matter, on the kinds of worlds they perform. These worlds or realities are not simply constructed; they have to be practiced or enacted into being, and this practice involves choices, obstructions, and interactions with other ontological realities. Although we are concerned to trace bottled water practices in a variety of settings, we are also interested in how these practices might interfere with other sorts of drinking practices. Interference here is an STS term that in Law's (2004a, 61) account means that realities are being practiced everywhere: they are complex, uncertain, and interact with each other—this is difference. These differences are where ontological realities may become ontological politics because difference can mean both conflict and dissent, the invention of alternative realities or their erosion. As Mol (2002, 7) says, "ontologies are brought into being, sustained, or allowed to wither away, in common, day-to-day, sociomaterial practices.... All of these, all at once, all intertwined, all in tension. If reality is multiple, it is also political."

A close study of bottles in action in Bangkok allows us to see these ontological politics in action. Our aim was to document the range of provisional arrangements that are in place for drinking water. These arrangements show how bottles are generating new practices, markets, meanings, and drinking performances. They are also being incorporated into existing practice-as-entity regimes in ways that complicate the distinction between product and service. The key purpose of this study was to investigate how the specific material affordances of plastic bottles, in all their varieties, are realized in practice, and how, in some arrangements, these practices enact ontological realities that interfere with the imagination and provision of more sustainable alternatives.

Provisional Water Arrangements in Bangkok Households

The process of organizing everyday drinking water involves collaboration with a range of human and nonhuman others in the household and beyond. In Thailand, rainwater was traditionally collected in large earthen jars placed outside households—a practice that persists in some parts of the countryside today. More recently, houses have been fitted with galvanized drainage gutters and metal pipes that direct the rainwater into large ceramic storage jars. In Bangkok and other urban centers, where atmospheric contamination of rainwater weighs against such practices, most households and condominium residents use the metropolitan tap water system, though in varied ways. Vestiges of past practices are nevertheless apparent among some of our Thai informants, who specified rain as a taste they valued in drinking water.

As we noted earlier, since 1999 the MWA has guaranteed that Bangkok tap water complies with WHO drinking water standards after treatment at the source. Water is regularly tested for key contaminants at various locations in the municipal network, and the results are uploaded to a consumer map on the MWA website—just one of many state-devised methods to enact "water quality" and establish public trust. Most of our informants were aware of state guarantees about water quality but expressed doubts as to the quality of pipes and their maintenance, especially in older buildings and sections of the city:

Actually in Bangkok they always promote that you can drink the tap water, but then we are worried, we know that the tap water from the factory, from the government's factory, is clean, but then, the tap itself, the pipes, we are not sure with that, what it goes through before it comes to our house. (Engineer, male, age forty-five)

Bottled water industry informants also promoted this view of pipes as an unreliable intermediary:

No matter how good your water is, somehow you have to find a way to deliver the water to the consumer and when you talk about tap water, you are talking about the piping system, and the piping ages—rust, pollution, all that sort of thing. People doubt the maintenance. (Industry representative, male, age fifty-one)

In these circumstances of perceived unreliability or risk, householders described entering into a variety of arrangements with mundane devices and technologies in the interests of organizing safe drinking practices. These arrangements included boiling water from the tap, installing filtering systems in kitchens, arranging home delivery of drinking water (home and office delivery, or HOD), or making use of the water-vending machines that can be found in some Bangkok neighborhoods.

In general, tap water was presumed to require special treatment, and this treatment took one of two forms, which often went hand in hand: grading and ranking water sources by using tap water for specific household tasks but not others, and using various methods to render tap water fit for drinking, traditionally through boiling, but also through other techniques, such as installing water filters on kitchen outlets:

I don't drink tap water at home. It's not clean. Sometimes they have something, when the water comes out. We use the tap water only for washing and shower. If you want to drink that water we have to boil it first. (Secretary, female, age thirty)

[Informant speaking of his friends and family] They use the tap water for taking a bath, cleaning, clothes or dishes. They use the tap water as a source some times for making tea. Generally it's an old habit of local people. If they want to drink tap water they usually boil it first, like if you want to make tea, if you want to make soup, use tap water. (Engineer, male, age forty-five)

At home I drink tap water that goes through this machine, just to get the germs out, installed at home, used for drinking. We just want to make sure that there are no germs in there. (Student, female, age twenty-four)

It's not convenient, but I've been boiling water since I was young. After you boil the water, we let it cool down, then we put it in the fridge. After you boil it all the smell is gone. These days, the trend in the house is they don't boil, they install the water filter in the house, for drinking water they use water from filter. If we have this bottle we collect it and fill it with filtered water. Then I will bring this to the school every day, filtered water. (Office worker, male, age forty)

As we can see, the grading and ranking of water from different sources is hardly a new feature of Thai households, where boiling water to render it fit for drinking (sometimes in the form of tea) has long been a familiar practice. In contemporary households, however, this practice is giving way to other arrangements, such as home treatment with water filters, for which the market was estimated to be worth 3.7 billion baht in 2012. Preceding this enthusiasm for home water filters were the coin-operated water vending machines that began to appear in some Thai neighborhoods in the late 1990s—an initiative of the Thaksin government (see figure 4.1). Essentially filtering devices, these machines were billed by the Thaksin government as a low-cost way to get safe drinking water from the municipal water system with the intervention of private operators.

I just moved from my old house. Before I used to use this machine, you put 2 baht and the machine gives you 2 liters, but the water is quite yellowish. So I decided to put the filter in at home. At first I thought, oh, it tastes better. (Student, female, age twenty-four)

Figure 4.1
Water vending machine in Bangkok.
Source: Kane Race.

The experiences with these public vending machines echoed many residents' experiences with or perceptions of the municipal water supply, insofar as they attributed deficiencies in the system to the ongoing problem of effective maintenance. The twenty-four-year-old student continued:

Before at Thaksin's government they used to advertise that these machines would cost about 800 baht for everyone to have clean water to drink. But then, as many as they got, nobody take care of it, nobody go clean the filter, so that's why it was like, aagh, yellowish.

Such shortcomings contributed to a general impression of state unreliability and reinforced distrust of the public provision of water fit to drink. Indeed, when one of our respondents was asked whether she expected the government to provide clean drinking water, or whether such a duty existed, she simply replied: "Look at what we have now, what kind of government!... I'm not sure. Let them take care of themselves first!"

In the context of these perceptions, it is not surprising that consumers sought assurance through other means, and the emergence of branded water—especially that produced by large established companies such as Singha or Nestlé—appeared to fulfill this role. Often consumers attributed the reliability of these sources to the quality of the machinery and the reputation of the brand. As commercial entities, the survival of these companies was thought to depend on adequate quality control procedures, and this was presumed to be an adequate measure of safety. Thus, large PET bottles bought from supermarkets gradually emerged as another domestic water source in greater Bangkok. Single-serve PET bottles, on the other hand, are generally not seen as a staple form of household consumption but are usually bought while out and about. They may, however, find their way into the house at the end of a daily excursion, where they are often refilled with boiled or filtered water and reused.

Each of these arrangements involves varying degrees of labor inside and outside the household, and the coordination of bodies, technologies, forms of competence, and routines. These routines are themselves entangled with other routines—preparing meals, going shopping, stocking the kitchen—and must be adjusted in relation to the material particularities and conditions of the elements at hand: water is heavy, transporting sufficient volumes for household use calls for a vehicle, Bangkok traffic is bad, the water-vending machine is hard to get to (see figure 4.2). What became apparent was the multiplicity of these practices and their adaptability to different conditions and elements over time. Indeed, across the urban land-scape of Bangkok, different methods for providing drinking water—and thus different water realities—frequently sit side by side (see figure 4.3), sometimes interfering with one another, at other times coexisting non-problematically, frequently vying for attention and putting themselves for-ward to invite different forms of practice and engagement with water.

One way of approaching this multiplicity of drinking water practices is to position the consumer as a participant in *provisional networks* of distribu-tion, preparation, and supply. In the household, these networks are provi-sional in two senses: they are about *making something available,* and they involve routinized practices that may nevertheless be subject to revision and some degree of change or innovation in the presence of new technolo-gies, services, and products such as filters, water-vending machines, or PET bottles. This metaphor of "making available" is particularly apt in the Thai context, as water is intimately tied in to practices of hospitality. For example, the first thing guests customarily are offered on entering someone's house is water. Indeed, a common expression for generosity is *nam chai*—literally, to have a water heart. The proliferation of new devices and mechanisms to provide good drinking water produces a dynamic relation between practice as entity and practice as performance. Drinking practices are also subject to change in the context of new styles of life, such as condominium living—a major and relatively recent development in Bangkok that, along with convenience stores, supermarkets, and shopping centers, is promoted as a hallmark of cosmopolitan modernity. Our informants were engaged—or had been engaged at particular points in time—in degrees of reflexive activ-ity about the various components of such networks in relation to everyday values such as reliability, cost, health risks, taste, social status, practical-ity, and convenience. This shows that it isn't simply new technologies or products that inaugurate changes in practices but rather their capacity to prompt reflexive activity and generate what we might call "ontological doubt" about the security and stability of existing arrangements.

Good or Service? Bottled Water Networks and the Relative Value of Disposability

In Bangkok, bottled water can thus be thought of as participating in a range of different water provision networks. One question that arises is

Figure 4.2
Delivering water in Bangkok traffic.
Source: Kane Race.

how certain qualities of bottled water emerge as significant in relation to the practices that make up Bangkok's drinking water realities. In Bangkok, home consumption of drinking water from nontap or non-MWA sources is largely dominated by the HOD industry and water from neighborhood vending machines—though use of PET bottles is on the rise. The HOD industry is popularly regarded as the environmentally friendly approach to dispensing bottled water because it reuses the large, nineteen-liter polycarbonate (PC) bottles in which it delivers the product. For this reason, HOD of drinking water is understood within the industry as a service:

The bottled water cooler industry differs from the bottled water industry that bottles water in small bottles in that the former is service-based, whereas the latter is product-based. This difference is very significant to the organizational structure of the bottled water cooler company. (Barnett 2008, 30)

Like other services, the provision of HOD water follows a logic that Callon, Méadel, and Rabeharisoa (2002, 208) describe as a "making available," in which the customer, "by opening a tap ... sets in motion a complex

Figure 4.3
Multiple water realities in a Bangkok restaurant.
Source: Kane Race.

arrangement of humans and non-humans whose actions have been adjusted in relation to one another and prepared for mobilization at any time and at any point of access to the network." Services imply an ongoing relation with the customer, in which the provider agrees to make certain things available on certain conditions for a period of time. Moreover, in service provision, customers become "an element in the system of action. They act, react and most importantly interact" (209)—a feature that Callon and colleagues see as productive of a customization of the relation.

As a service, the organizational structure of the bottled water cooler company emerges in part from the material specificities of the objects with which it deals. These specificities give rise to practices that inform the organizational structure of the service provider. In the case of home-delivered bottled water, because of the nature of the service, the company must arrange regular scheduled delivery and collection, maintenance and repair of the relevant technologies, customer care, and industrial cleaning and reuse of the nineteen-liter PC bottles in which water is transported and dispensed. Cleaning the PC bottles is a complex but necessary operation that involves applying cleaning solution and rinsing the inside and outside of the bottles using a jet-spray system. This operation must also respond to customer activity, because in service arrangements, customers become an "element in the system of action" (Callon, Méadel, and Rabeharisoa 2002, 209). As one HOD industry informant explained:

When your bottle comes from the market, you never know what your bottle has been in contact with, what kind of contamination. That's why our company tried to persuade the consumer to recap your bottles after use, but we haven't been success-ful as yet, because people say, what the hell? You know, so the contamination comes along. The worst part is when your customer or your consumer doesn't really, you know, care about, take care of the bottle. They put it as a container for some kind of petroleum, or some kind of pesticides, or whatever, and when you bring this bottle back, the only solution for that is that you have to discard the bottles. So when the bottle comes back from the market the first thing that the operator does is, they have to do the sniffing test. And if it smells of petroleum or something then they put it aside. The majority of times you have to destroy the bottle. And then they do the cleaning of the outside, which is quite a difficult one, and after that, you do the cleaning of the inside. (Industry representative, male, age fifty-one)

Here, the industry representative positions customers as "unreliable intermediaries," and we can see how the company's reuse of the bottle requires it to account for the customer–bottle relation. In other words, the reusability of the bottle necessitates certain relations among providers, cus-tomers, and bottles, which renders each of these entities an active element in the network of provision.

Alongside their theorization of services, Callon and colleagues (2002) cast *goods* in terms of the sequence of actions and operations in which their properties are worked on and qualified. Here, products are goods seen from the point of view of their production, consumption, and circulation—a process that involves various forms of organization and reflexive activity on the part of economic agents. These processes include specific devices for registering and incorporating consumer desires and preferences, which in turn inform how the good is worked on, qualified, and positioned in the market. Of note, the qualification of a product can consist of work on the image of the product, or work on its actual material form, or both. Callon and colleagues argue that there is no need to distinguish between primary and secondary qualities in this regard: for the purpose of market position-ing, these attributes share the same ontological status. This is a valuable framework for thinking about how the properties of goods are continually modified in interactions with the economic agent (or customer) and other elements. It allows us to appreciate how the quality of disposability associated with PET bottled water might emerge as particularly significant within this relational network. Disposability is a quality of PET water goods that distinguishes these goods from services based on the reuse of PC bottles. In this respect, we can see what becomes available to producers in this network of relations: the ability to opportunistically position PET bottles as a solution to the shortcomings of some of the prevailing service arrangements for accessing drinking water, namely, the difficulty of identifying and excluding unclean bottles in the HOD system.

To think further about this quality of disposability and how it is constituted in relation to bottled water, it may be useful to consider how it operates and acquires value in other contexts. Apart from the HOD industry, another example in which bottles are reused in the Thai context is with single-serve glass bottles, once commonly used to sell soft drinks. (In Thailand, traditional trade channels actually prefer this option because of the higher margins involved.) We can see that in both these cases, reuse of the bottle makes certain demands on companies and distributors, ranging from the need for storage space in the case of empty glass bottles (which makes modern and convenience stores reluctant to take up this option; see figure 4.4) to the pickup and industrial cleaning of used bottles in the case of HOD services. Moreover, the premium on the bottle's reusability grounds the bottle in certain ways. If the bottle travels too far, the distributor experiences a loss. Thus the vendor must account for the mobility of the bottle. This is usually achieved by restricting distribution to certain locations (such as the home or office in the case of HOD services), by introducing a deposit/refund system, by requiring consumption on the spot (for example, in restaurants or at food stalls), or by other creative means. For example, soft drink vendors working with glass bottles in markets often transfer the contents to a plastic bag, add a straw and ice, and fasten the bag at the top with a rubber band, allowing the vendor to retain the bottle while the consumer is free to roam with a drink in hand.

In these instances, we can see how a property of the bottle (reusability) institutes specific relations and practices and necessitates certain sociotechnical arrangements on the part of suppliers, distributors, and consumers. These examples also help to reveal some of the specific affordances of the

Figure 4.4
Storing bottles: multiple trade channels.
Source: Kane Race.

disposable PET plastic bottle. In quite material and practical ways, the single-serve PET bottle functions to equip or constitute the mobile subject of self-hydration. Disposability allows a severing of the relation between the provider and the consumer. Indeed, the single-serve PET bottle was most often associated with consumption on the move, rather than in the household, the main appeal being its convenience. But what can we say about the growing use of PET bottles inside the home? Since mobility is not a salient value in this context, how does this product position itself in relation to the alternatives?

It is precisely the material performances of HOD networks—and their tendency to fail—which the PET bottled water industry seizes on in its attempts to penetrate the household market. As the head of bottled water at Nestlé Thailand reported in the region's industry magazine:

The new generation of consumers are health-conscious so they started questioning the cleanliness of the returnable 19-liter bottles, especially when the bottle condition and label looked old, alerting a concern on the potentials of poor washing and disinfection. (Muernmart 2008, 8)

Here, the company makes reference to customer experiences of HOD bottles, in particular the mishaps associated with the reuse of bottles in given networks of supply. The property of disposability acquires value in relation to bad experiences with the reusable bottles that characterize the household market. It was precisely such observations that prompted Nestlé to introduce a new six-liter PET bottle into the modern convenience store trade channel. And what is especially revealing here is that the positioning of bottled water within this market involves specification of not only the symbolic but also the material properties of the product. Consider, for example, the material-practical concerns that are cited to position it:

Convenient to buy, not that heavy, suitable to carry back home, and a good price per liter—ideal for the new generation family of 2, husband and wife. (Muernmart 2008, 7)

Of note in this passage is not only the allusion to a new, "modernized" consumer with a modern lifestyle but also the company's familiarity with the provisional networks in which such consumers participate. Once again, the bottle—including its actual material form—is carefully adjusted to, and designed to enable, certain routines of modern life. The appeal to new styles of life is not merely symbolic but promotes the six-liter bottle in terms of its specific affordances: not too heavy, but containing a sufficient quantity of water for a modern household; something you might pick up at a convenience store on your way back to the condominium. And we note also that these material properties gain their value and significance in relation to existing provisional networks, specifically concerning insecurities about the cleaning of bottles within HOD arrangements, and thus the quality and possible contamination of the water they contain.

Conclusion

What can we take from this study of drinking water in Bangkok? Perhaps most obviously, we see a close attention to consumer practices, routines, and provisional arrangements on the part of bottled water companies. Consumer perceptions of tap water were affected by the appearance of new technologies and products that promoted themselves in terms of safety and reliability. Most of the provisional arrangements we encountered were driven, at some level, by concerns about safety—though it was often difficult to distinguish safety concerns from other concerns. In these contexts, tap water was relegated to use in other, nondrinking household activities, such as cleaning, bathing, and cooking. Ontological doubt about the safety of different provisional arrangements merged with other rationales for consumption, including social status, the taste of the water, convenience, and health. Throughout this context, companies pay close attention to actual drinking arrangements and the ways in which they continually qualify and requalify water. The promotional appeal to a modernized lifestyle goes hand in hand with a positioning of products that emphasizes their material properties and affordances as much as their image. Indeed, these material dimensions become part of the brand, insofar as the bottle is positioned and experienced in terms of everyday values such as practicality, convenience, and cleanliness.

Second, the material properties and affordances of the bottle acquire their value in relation to the provisional arrangements to which households are accustomed, and their perceived shortcomings. The quality of disposability afforded by PET plastic has no real practical value in the household except in relation to some of the undesirable aspects of existing provisional networks, such as rusty pipes in the public water supply or the cleaning practices of the HOD industry. The salience of these qualities depends on, or contrasts with, the performance of other materials against which it is compared, such as reusable PC bottles. In this respect, materials such as PET plastic can be regarded as involved in "overlapping webs of relational performance," to borrow Shove and colleagues' (2007) suggestive phrase. These performances consist of specific applications of given materials and are themselves "relative, provisional and inherently precarious" (105). Producers seek to make new markets by promoting specific expectations of material object performance. Meanwhile, the material value of PET plastic takes shape in relation to given performances of PC plastic.

This point connects with the next observation we can draw from the Bangkok example, which concerns the distinction between goods and services. Indeed, when it comes to water provision, it is tempting to argue that the quality of disposability enacts a distinction between product and service in the home space. If within the logic of service provision customers are constituted as active elements within a system of action, we can see that much of the appeal of PET plastic is that it severs the need for continued participation in the network, or indeed any ongoing relation between customer and provider. This value can be understood in terms of mobility (that is, dispensing with the need to return the bottle to the provider) or in terms of dislocation (releasing the customer from a provisional network in which certain elements, such as dirty bottles, are unsatisfactorily controlled or accounted for, thus creating access to options for safer or cleaner water). In either case, with a disposable PET bottle, the ongoing relation between customer and provider can cease at the point of purchase. A question that arises here in relation to the circulation of a good so basic to human sustenance as water is whether severability from provisional networks (as depicted in the logic of products and services) is an adequate formulation, or whether the provision of clean drinking water ought to be characterized in other terms entirely—not in terms of the logic of choice, for example, but a logic of care, which Annemarie Mol (2008) characterizes as an interactive, ongoing, and open-ended process that does not stop at the point of transaction but rather requires continual modification, depending on results and human need.

Finally, the multiplicity of drinking water arrangements in Bangkok, and their interaction and interference with one another, suggests the need to further theorize *convenience*. Not only is convenience commonly cited to account for consumer preferences, it is frequently cited specifically in relation to bottled water marketing (Ward et al. 2009). Elizabeth Shove (2003) discusses convenience in terms of the ability to shift time. In the case of bottled water, we can see that it is also connected to practices of movement, and thus has spatial dimensions. Etymologically, convenience implies a coming together. It is a coming together of different elements in a network of humans and nonhumans in an arrangement that is adjusted to the routines of key actors in that network. As a demand, it takes its bearings from given routines, procedures, and competencies. Just as often, though, it is projected as a property onto specific goods, services, or arrangements. Convenient products are those that are well attuned to stabilized routines and procedures in given relations of affordance. Or they promise a new, more desirable or efficient stabilization, which allows the redistribution of specific forms of labor, cost, or time. But what is also apparent from our study of Bangkok drinking practices is that convenience is not a transparent value. As we have discussed, it takes its bearings from given configurations and provisional networks. In what situation is lugging a six-liter PET bottle home from the convenience store convenient? As well as citing and differentiating itself from existing routines, the quality of convenience disrupts or supplants them, making certain practices and competences redundant, and creating the need for new forms of labor, cost, and routine. This is how we understand Mol's (2002) and Law's (2004b) claims that different practical ontologies interfere with one another. The value of the social practices approach we are advocating here is that it can better take account of the dense materiality of stuff like PET plastic and the subtle ways it works itself into everyday lives, while also exposing the contingency of these workings, their relationality. And this in turn might make emerging bottled water drinking practices more open to change, less "normal."

5   Enacting Water Scarcity in Chennai

This chapter develops our discussion of bottled water's role in the emergence of new drinking water practices and realities. The primary focus is on the imbrication of bottled water markets in growing water scarcity in the city of Chennai, the capital of Tamil Nadu and India's sixth largest city. Although bottled water initially appears to be the solution to inadequate and unsafe water provision in this city—commodity abundance in response to public scarcity—we want to query this teleology by investigating exactly how bottled water production and the markets for bottled water are connected to wider and extreme shortages of water. Rather than characterize bottles as an effect of or solution to a very serious water reality, our aim is to understand them as participants in the emergence of this reality. The issue concerns the capacities and effects of bottled water in the context of a declining supply of water. How does a market interact with wider biophysical and biopolitical water realities? In what ways is it implicated in bringing these realities about, not just responding to them? And how does the "becoming normal" of bottled water render scarcity a problem, particularly for those who cannot afford to opt out of inadequate urban provision?

As a result of the water challenges facing Chennai, the consumption of bottled water has risen dramatically in recent years. This increase has occurred in parallel with broader patterns of market growth across India, a country that has long struggled to provide adequate safe public water to its population. Between 2000 and 2005, per capita consumption rates for bottled water tripled (Arnold 2006). Among public water advocates in India, bottled water is often represented as a wasteful folly and its consumption as indicative of a growing middle class seeking to circumvent inadequate public water systems rather than advocate for their improvement (Raja 2008; Shiva 2002). This assessment fails to take bottled water seriously as an active participant in the production of contemporary water realities. In Chennai, bottled water asserts itself in ways that demand a different approach. Here, the story of bottled water is a story of water scarcity in the making, as a particular cluster of factors has created opportunities for markets to rapidly grow. As more and more people turn to bottled water, it is becoming a key player in existing water regimes and changing them in significant ways. These water regimes include inadequate or absent public water infrastructure in many areas, a history of overdrawn local groundwater sources, and a growing popular acceptance of water scarcity as a normal reality. For those who can afford it, bottled water emerges in this context as more reliable and accessible than the other options available. But this is not just a story about bottled water consumption as a response to scarcity. In Chennai, the production, transport, and sale of bottled water for local and national markets have helped create the very conditions of water scarcity that the product is employed to address, just as in Bangkok bottled water is intimately implicated in the shifting dynamics of the availability of potable water throughout the city and its environs. However, the specific hydrological history of Chennai means that the bottle enacts quite different realities and practices.

Inadequate public water systems have long been a problem throughout India. As Matthew Gandy (2008) has argued, India has never experienced a "water modernity," despite its postindependence investment in large-scale infrastructure projects such as dams and river diversions. In this process of "development," precolonial practices of localized water management became unworkable in urban contexts. According to Gandy, the "rationalized visions of urban space" brought by colonial influence failed to match the multiple, inequitable modes of dwelling that flourished in India's booming postindependence cities (108). Nor did the precolonial practices cater well to the seasonal monsoons that brought rainfall in uneven bursts across the country (126). In this context, the capacity to negotiate the poor drinking water endemic to India's large cities tended to be economically patterned. Although a deficient public system was technically nondiscriminatory, affecting all equally, those residents with greater material means had the freedom to "opt out" of an ongoing struggle with degraded public water by purchasing their own private water infrastructure instead (121). Gandy does not include bottled water in his discussion of the rise of opting-out strategies, but its rapidly growing presence in India's urban water landscape fits his argument: packaged water manifests as one of the key material means by which wealthier residents have been able to bypass inadequate public systems and arrange water fit for drinking on a daily basis.

What makes the Chennai case so compelling, however, is its complication of the assumption that choice and privilege entirely motivate the consumption of bottled water. Bottled water has entered into normalized routines of water provision in Chennai in such a way that its assumed availability and reliability are now embedded in patterns of urban organization. Housing development has boomed in Chennai in recent years, particularly on its urban fringe, despite inadequate planning rules relating to the provision of public water supply. This urban development is viable only because of available bottled water arrangements (Janakarajan et al. 2007, 58–59). This means that Chennai's relationship with bottled water is in many ways unique. Bottled water plays a much more noticeable part in daily domestic life for a significant portion of Chennai's residents than the other "opt-out" strategies favored in other Indian cities, such as installation of a home water filter and storage systems. Indeed, it is very much connected to the failings of the public water supply and the emergence of water scarcity in this location, as we go on to explore.

Explaining Water Scarcity

Before we embark on a detailed investigation of bottled water in Chennai and its implication in growing water scarcity, it is worthwhile briefly examining wider debates in water policy and development discourse about this disturbing reality. In India, where the availability of fresh water has declined dramatically since the mid-twentieth century (Ray 2008, xi), these debates have tended to focus on the management or mismanagement of water resources. Antimarket commentators argue that water scarcity is the result of India's postindependence embrace of modernization and the market economy; they point to the growth of water-intensive industries, the marginalization of traditional water knowledge and practices, and the degradation and pollution of water environments. For example, the water activist Vandana Shiva writes, "[In India], the story of water scarcity has been a story of greed, of careless technologies, and of taking more than nature can replenish and clean up" (Shiva 2002, 2). Pro-market commentators, on the other hand, tend to assign responsibility to "subsidized (state-managed) water" for leaving many of "the poorest of society without access to good clean water" (Landry 2002, 2). This literature routinely names government inefficiencies, centralized control, out-of-date infrastructure, and the general dysfunction of "the state water machinery" as the major factors responsible for water scarcity (Briscoe, Malik, and World Bank 2006, 4). While these accounts recognize the role of various sociopolitical factors, they also tend to place water scarcity in what Simone Abram and Marianne Elisabeth Lien (2011, 3) refer to as the "nature category." That is, water scarcity is essentially viewed as an always possible state of nature brought about by inappropriate practices and variable rainfall. This "naturalizing" tactic produces "scarcity"—and its corollary, "abundance"—as ahistorical and generalizable concepts. It also positions water scarcity as a largely biophysical condition that can emerge with predetermined meanings and effects across diverse locations.

In contrast to this naturalizing tendency, recent water scholarship has investigated "scarcity" as a constructed reality discursively employed to justify particular state and market water practices. Speaking in the Indian context, Lyla Mehta (2007) explains that the declaration of a region or town as "water scarce" is frequently given as the justification for large-scale interventions into the distribution of water, often through public-private partnerships that favor particular economic and political actors (655). In a similar vein, Karen Bakker (2010) and Maria Kaika (2006) argue that "scarcity" is a relational and contingent concept rather than an absolute. According to Kaika, it is only within a particular "social and political nexus" that scarcity has meaning (162).

From this perspective, water scarcity is not a predominantly natural condition but a reality that is constituted and mediated. It is, we could say, materially enacted in practice through a range of interests and actors. But how exactly does this happen? In what ways is scarcity enacted? In answering these questions, we will see how the enactment of water scarcity in Chennai works to normalize bottled water as an everyday water source and how, reflexively, this normalization of the bottle helps to enact scarcity. While not a dominant technology of water delivery, bottled water is nonetheless pervasive, highly visible, widely distributed, and increasingly consumed as part of daily water regimes for residents. For many, it has become a significant means of obtaining water for subsistence purposes. Today, it is estimated that one-third of Chennai's population uses bottled water in a range of forms on a daily basis (Mariappan 2011). In contrast to the city's inefficient and inadequate state-supplied water network, bottled water represents on-demand water availability, a kind of commercial water abundance, unlimited and reliable—for a price. However, this abundance is highly contingent: it depends on certain forms of water provision made available in the context of specific market relations. In this way, bottled water emerges as a dependable and convenient water source when other modes of water distribution have failed.

A History of Scarcity in Chennai

Chennai is well known for chronic water shortages. Despite an average rainfall greater than that elsewhere in India,[1] the per capita availability of potable water in this city is currently rated one of India's lowest (Gopakumar 2012, 59). As a result, the city has been called the country's "water scarcity capital" (Desai 2010), a designation that reflects a history of poor infrastructure, regulatory failure, and steady groundwater decline. Until the 1970s, the city still relied on its four main reservoirs to supply its public system. However, the increased use of water for agriculture and a burgeoning population in the second half of the twentieth century (the population quadrupled in size between 1960 and 2000) saw these reservoirs come under increased stress as their per capita availability dropped by up to 40 percent (Vaidyanathan and Saravanan 2001, 4). In response, the Chennai Metropolitan Water Supply Board, an autonomous statutory body charged with the provision of water to the city's residents (popularly referred to as "Metro Water"), commissioned the sinking of new wells to access groundwater in a widening radius around the city in a bid to supplement surface water with groundwater supply. Reliance on groundwater has continued to grow in Chennai, and, as of 2012, this source provided 60 percent of the city's water (Gopakumar 2012, 62).

This exploitation of groundwater has been prevalent throughout postindependence India. It was connected to the pursuit of economic development and the water-intensive requirements of the green revolution. In the 1970s, the World Bank began to subsidize mechanized groundwater withdrawal systems, including the provision of free electricity to irrigators, and provided credit for the installation of tube wells to enable large-scale irrigation (Briscoe, Malik, and World Bank 2006, 22; Shiva 2002, 10). Tube wells—long stainless steel pipes that draw water up from the ground by electric pump—are so efficient that groundwater can be withdrawn much faster than it is replenished. More than 20 million tube wells have now been installed throughout India, with 50 percent of agricultural land irrigated by this technology (Barlow 2007, 13). This has resulted in the widespread over-drawing of aquifers and a subsequent trade in water between agricultural lands. Feeding into this picture is the increased pollution of surface water throughout India with industrial and human waste. Depleted groundwater resources cannot recharge surface water, leading to its further degradation, leaving groundwater the only source of viable potable water. As a result, groundwater sources are at unprecedented lows throughout India. While nationally, 15 percent of aquifers are in a critical condition, Tamil Nadu has the most compromised aquifers in southern India, with approximately 40 percent of its groundwater currently overexploited (Briscoe, Malik, and World Bank 2006, 25).

A lax regulatory climate facilitated India's investment in groundwater, and for much of the twentieth century, its extraction went largely unmonitored. In Tamil Nadu, it was not until 1987 that the Madras Metropolitan Area Groundwater Regulation Act was introduced to prevent the indiscriminate access and use of groundwater in the Chennai area through the allocation of permits intended to limit the volume of water taken from aquifers. This was followed in 2003 by the Groundwater Development and Management Act, which gave the Tamil Nadu Groundwater Authority the power to regulate and control water development in the state. The impact of these regulatory efforts has been compromised by the state government's failure to enforce the 2003 act. In February 2011, the Madras High Court issued a directive to the Tamil Nadu government to refuse any further extraction and sale of groundwater until the act's notification; however, the precedent for sidestepping such legal restrictions is strong, and it is commonly understood that violators of the 1987 act are rarely pursued (Potter 2009a). The permit system has also been criticized for its failure to coherently track or maintain data on volumes extracted.

This regulatory failure has encouraged a culture of "self-sourcing," which some defend as the only viable option for individuals and industries faced with an inadequate or restrictive public supply. According to a report written for the World Bank, "self-provision is the best indicator of the failure of public water supply systems. Tubewells proliferate ... because public irrigation managers are unable to deliver irrigation on demand. Urban households want their own boreholes because the municipal service is inadequate and unreliable" (Briscoe, Malik, and World Bank 2006, 23).

By the mid-2000s, poor monsoon seasons coupled with the accumulating effects of groundwater extraction over the previous decade had intensified Chennai's reliance on groundwater. Sample testing across the city in 2003 revealed the degraded quality of its aquifers, which were high in total dissolved solids, salinity, fluoride, calcium, and magnesium. These levels were well above those classified as "good quality": by 2004, 67 percent of the city's groundwater was considered to be of "moderate quality," while 33 percent was classed as "bad quality" (Lakshmi 2009). In the preceding ten years, Chennai's groundwater had declined by four meters, to below sea level, allowing saltwater intrusion into many coastal aquifers. The city's reservoir levels correspondingly dropped, domestic wells started to dry up, and Metro Water responded by furthering its radius of groundwater extraction, sinking new boreholes fitted with hand pumps around the city and introducing rationing systems, such as alternate-day or partial-day supply. By 2005, Metro Water could meet just over one-third of daily demand in the greater Chennai area as a result of its intrinsic and extrinsic limitations (Narain 2005, 5). Moreover, Metro Water is responsible only for servicing metropolitan Chennai. The rapid expansion of urban development on the periphery of the city outside the designated area of piped supply—including on land that has been depleted of groundwater in a bid to supply the inner city's water needs—means that many local residents do not have access to any sort of regular public supply.

But even for those "on the network," access to water is not the only concern. Undrinkable and unsanitary water in the public system is an ongoing issue: as groundwater is sourced from lower and lower depths, the risks of contamination increase, making piped water often taste and smell unsavory. Metro Water's pipes are also vulnerable to tapping by individuals with booster pumps, and this siphoning off further compromises what flow there is. Meanwhile, pipes servicing multistory apartment buildings often have inadequate pressure. The routine practice in middle-class households is to use tap water for bathing and domestic cleaning only, and bottled water for drinking, and often for cooking too (Potter 2009b). Tankered water delivered in trucks is also part of the Metro Water system and is equally untrustworthy. In 2009 the residents of the Chennai suburb

Virugambakkam reported that their boreholes filled from Metro Water tankers had turned black and had an odor of sewage; several children fell ill with typhoid and malaria. Newspaper reports of this event recounted local doctors prescribing premium bottled water in response (Sujatha 2009). The groundwater transported by Metro Water is minimally treated with ozone and perishes after ten days or thereabouts, when worms often start to appear in the water. The regularity of supply is also an issue. In 2010 the suburb of Anakaputhur, for example, reportedly received water from Metro Water's tanker trucks only once every fifteen days (*Deccan Chronicle* 2010b).

Illegal activity and corruption have also been implicated in the uncertainty of public supply. A handful of local politicians reportedly own private water tankers hired out to Metro Water, enabling their own interests to be inserted into the process of water supply. Metro Water representatives have also been accused of demanding money for "releasing water" to particular suburbs over others (Frederick 2003). But the logistics of local water delivery also prove complex and unwieldy for a large bureaucratic entity like Metro Water to handle: not all areas are accessible to industrial-size tanker trucks, while delivery schedules are frequently unreliable. A local paper reported on truck drivers refusing to service certain areas on certain days because it was not convenient (*Deccan Chronicle* 2010a). Metro Water is said to receive 15,000 complaints from its consumers each year, and introduced an SMS facility to deal with this volume (*Times of India* 2010).

Making Water Abundant: The Role of the Bottle

As these diverse factors materially contributed to the emergence of water scarcity, an alternative reality of water abundance was simultaneously enacted. Tamil Nadu was home to almost half of India's 1,400 bottled water producers in 2011, with a recorded 270 water bottlers in Chennai alone (Mariappan 2011). Most of this water is derived exclusively from groundwater sources. In 2005, 700,000 liters of bottled water were sold in Chennai every day, accounting for 25 percent of the national bottled water industry revenue (Murthy 2005). Over half of Chennai's bottled water market consists of "bulk" twenty-liter containers, transported and sold to households and commercial establishments by a host of distribution companies, much as in Bangkok. In this sense, bottled water industries participate significantly in the exploitation of groundwater and overdrawn aquifers in this region.

Until the 1990s, bottled water was an obscure and largely boutique product in India. However, with economic liberalization and the growth of PET packaging industries, market development and product diversification expanded dramatically. These changes saw a shift in the kind of water sold, from "mineral water" aimed at high-end consumers to what is known in India as "packaged water," meaning minimally treated groundwater (and, less commonly, surface water) distributed in a variety of package volumes, commonly twenty-liter, one-liter, 500-milliliter and 200-milliliter containers. These products sought to extend the bottled water market far beyond the luxury segment and, in the words of the founder of leading national brand Bisleri, "reach out to the masses" (Raturi 2005). By 2010, the Indian bottled water industry was worth U.S. $250 million (Rico 2011). This growth, estimated at three times the overall rate of Indian economic growth in 2008 (Raja 2008), was enabled by the accessibility of groundwater resources, in addition to the relatively low industry standards imposed by government regulators. Licenses to manufacture bottled water were particularly easy to obtain until the early 2000s, when, in the wake of widely publicized contamination scares, the government narrowed its definition of what constituted "purified" water (Bhushan 2002). In 2007, thirty-two Chennai-produced packaged water brands were declared unfit for human consumption, and nine production units closed as a result (Nadu 2007). Yet licenses are still widely allocated: more than 150 new licenses were issued in Chennai in 2010 alone (Mariappan 2011). A significant number of unlicensed bottled water manufacturers also continue to operate throughout India, including in Chennai.

The private extraction of groundwater for bottled water manufacture has also benefited from the longer history and normalization of water distribution technologies that involve containerization, particularly the increasing use of tanker trucks in the dissemination of public water and the transport of privately commissioned bulk water across the city. Amid these practices, bottled water lends itself easily to distribution by a network of actors distributing water outside the pipe and the tap (see figure 5.1).

The logistical and technical problems Metro Water faces in providing a reliable and efficient water supply have meant that the opportunities for small-scale bottled water providers have flourished. Water vendors service Chennai neighborhoods on motorbike; one vendor profiled by the local paper uses a missed call system on his mobile phone to communicate with

Figure 5.1
A home delivery water vendor in Chennai.
Source: Emily Potter.

his customers. The door numbers of his customers' homes are stored on his phone; when a customer calls twice, he knows that he or she requires water, and he will promptly deliver to the customer's door (Varma 2009). Systems of convenience such as this promote bottled water consumption as not just safe but also dependable—always available, any time of the day or night. Bisleri and other brand-conscious bottled water manufacturers also offer a twenty-four-hour convenience home delivery service: "Picture this," an advertisement for this service explains. "It's the rainy season. You come home late, tired and discover that your building doesn't have electricity and water and you have run out of drinking water as well.... Just pick up your phone and dial Bisleri" (Dasgupta 2010).

This service caters to the upper end of the market, of course, indicating the elite status that certain brands of bottled water carry in India. However, the lower to middle end of the bottled water market has also flourished in Chennai, owing to its increased affordability and availability. According to one informant, a fifty-eight-year-old activist and physicist who has sought to address water shortages through the local practice of rainwater harvesting in his apartment complex, bottled water use has become so routine and familiar for many residents that even when alternative and more sustainable sources are readily available, bottled water is still preferred: "It is an addiction," he said (Potter 2009b). For Chennai's poorest residents, the cost of much bottled water is prohibitive. However, even for this demographic—often the most likely to be without a networked supply—water vending in other forms plays a central role in meeting daily water needs. This often involves local water kiosks and private taps where people can refill plastic bottles for a price. Different types of plastic are used for different bottled products, with the cheaper end of the range utilizing thinner, more porous material. For example, 200-milliliter bottles—or "cups," as they are called—are sold cheaply, for a few rupees, often in high-traffic contexts such as railway and bus terminals.

While the experience and practices of bottled water consumption vary across economic groups—there is obviously a significant difference between home delivery of a premium brand and filling your own bottle from a neighbor's tap—the value of containerized water is meaningful only in relation to an unreliable public network. The idea of simply picking up the phone and dialing Bisleri, or of sending a text to a local water vendor, creates a reality in which water is rapidly delivered in response to demand, in which the market appears as a far superior form of supply. At the same time, the ongoing extraction of groundwater in Chennai's environs means that bottled water is inextricably caught up in advancing water scarcity even as it positions itself to consumers as a reliable salve to the ills of a dysfunctional piped network and drought-stricken environment. After all, this is India's water scarcity capital *and* the epicenter of the country's water containerization industry.

Bottled water's capacity to generate alternative water realities in the midst of water shortages thus relies on its ongoing participation in the enactment of scarcity. It does this not only through the literal extraction of groundwater but also through discursive participation in the framing of scarcity as a justification for reaching for bottles, and through its efforts, by means of marketing and positioning strategies, to render scarcity highly visible. The rapidly increasing and normalized presence of bottled water on the streets on Chennai—whether for sale at roadside stalls, being transported on the backs of vans, or in the form of empty bottles sitting by the front doors of countless homes and businesses—has become a sign of scarcity itself. As the same fifty-eight-year-old informant told us, the increased visibility of bottles and tanker trucks on the streets of Chennai always heralds a "bad" summer to come.

This specific economy of qualities for bottled water has been a deliberate strategy undertaken by the industry across India generally. Bottled water was initially marketed in India as an aspirational lifestyle choice and a healthier beverage option than soft drinks. It was not just sugar concerns that drove this market: in 2003, a scare over toxic traces of metals (particularly cadmium) found in Indian Coca-Cola products resulted in a bottled water boom. Where street vendors once displayed piles of Coke cans, bottled water began appearing (Potter 2009c). But the lifestyle marketing of bottled water branding also emphasized its elite status, and endorsements from glamorous Bollywood celebrities were sought for this purpose. Reflecting this, the newspaper *India Today* in 2001 proclaimed bottled water to be "wet and sexy" (Renuka 2001). By the early 2000s, however, a different pitch had also entered the marketing of Indian bottled water, and with it the industry took off, reaching annual growth levels of 40 to 50 percent by 2005 (Murthy 2005). This pitch played on concerns about scarcity. The chairman of Bisleri explained his product's success in these terms:

There was a clear opportunity for building a market for bottled water. The quality of water available in the country was bad. It was similar to what Europe faced before World War II. The quality of water in Europe was extremely poor, which created a bottled water industry there. In India, too, not only was water scarce, whatever was available was of bad quality. (Raturi 2005)

Bisleri's 2001 catchphrase, "Play Safe," activated these associations; similarly, the Coca-Cola–owned brand Kinley's 2002 slogan *boond boond mein vishwas* translates to "reliability in every drop" (Murthy 2005). These reliability claims for bottled water reference not just the quality of the water itself but also its availability.

This marketing language tapped into a local and national discourse of water scarcity. During the "drought" of 2003–2004, *The Hindu* described Chennai as "thirsting for water" and "facing acute water scarcity" (Frederick 2003), while in 2002, an online activist/media site, the India Resource Centre, declared that "water scarcity in the metropolis reaches crisis proportions every summer." Its coterminous survey of the country's booming bottled water market was headed: "Water Business Thrives amidst Scarcity"

(Kamat 2002). Metro Water was widely condemned for failing to respond to the rapidly unfolding problem of water scarcity. As a commentator on the matter put it, Metro Water was "so removed from the symptoms of water scarcity [that it] hardly fathom[ed] the extent of the water troubles faced by Chennai's citizens every day" (Narain 2005, 4). The 2002 Indian National Water Policy also tapped into this discourse, flagging water as a "scarce national resource to be planned, developed, conserved and managed as such" (Lahiri-Dutt 2008, 7).

This strategy of addressing scarcity with reliability and presence has also been relevant in the context of the health concerns that have plagued India's bottled water industry. In the early 2000s, the issue of contaminated bottled water became highly visible as random sampling undertaken in Delhi revealed high traces of contaminants in all nationally bottled brands—something the tightening of license allocations was intended to address. In Chennai, the repeated reuse of containers by bottlers is a particular factor in ongoing concerns with water quality in twenty-liter bottles, as is also the case in Bangkok. Residents have reported falling ill with throat and respiratory infections after consuming particular brands (Lakshmi 2005). Under the logic of consumption, however, it is easy enough to move on to a different brand, given the proliferation of bottlers. Another strategy that has emerged is boiling lower-end bottled water before consuming it (Velloor 2005). Certain brands are trusted above others and purchased by different demographics, enabling bottled water to continue in its generalized enactment as a solution to scarcity and an ever-present agent of reliable water delivery. The market context frames untrustworthy bottled water as a fault of individual producers and imprudent or unlucky consumers rather than a matter of wider water realities.

The bottle's constant performance of abundance, even as it renders scarcity visible and material, keeps these multiple realities in constant tension. The opportunity for regular supply afforded by the bottle makes new water realities possible; it also makes these realities self-sustaining. What we mean by this is that, as bottled water allows the ongoing development of the city despite its seriously compromised groundwater resources, it also ensures an ongoing demand for bottled water, as these communities increasingly depend on bottles in daily potable water regimes. Twenty-four-hour delivery services at the other end of a phone call, well-stocked local shops, and tankers that transport thousands of liters of drinking water on a daily basis into water-scarce Chennai have qualified bottled water as reliable, convenient, and, most important, *available*. At the same time, this availability—a new kind of commercial water abundance, in contrast to a tap with no water—feeds directly into the challenge faced by many residents of the city, though generally the poorest, in accessing drinkable water. These multiple realities mean that the presence of bottled water in Chennai's water landscape is diverse and contradictory. In a variety of ways, then, bottled water is an undeniably active force caught up both in the making of scarcity and in its redress.

Conclusion

Although it is often positioned as a luxury, bottled water in Chennai is increasingly an essential good, entering into the common infrastructure of daily water access. This infrastructure is not horizontally accessed but is stratified by economic means, and this differential accessibility has consequences for how life is conducted across the demographic spectrum. The communal tap in a poor neighborhood generates a very different water reality from that of bottled water in a middle-class apartment building. The former may provide water access, but this access is often irregular, the water may be unsafe, and the various practices engaged in to obtain water from the tap may involve queuing for lengthy periods, thus shaping other domestic, social, and economic possibilities for individuals, especially women, and families. If several hours per week are needed to obtain water, other kinds of labor must be deferred during this time. Under current conditions, bottled water is a cost that many residents are prepared to pay in order to avoid the irregularities and inconvenience of other arrangements. Scarcity and abundance can thus coexist, even while the possibility of flipping from the latter condition to the former is ever present—after all, the supply of groundwater in the Chennai region is not unlimited.

This reflects Mol's (2002, viii) point that enactments "come in the plural," which refers to the multiplicity of practices that bring realities into being. In the case of water scarcity in Chennai, these multiple enactments challenge the notion that the water crisis is merely the end-point of an inadequate public supply, historical decisions, political interests, market opportunity, or discursive practice. Rather, it manifests in the active unfolding of these things, and through all the more-than-human materials, actors, and relations that constitute their enactment. Far from a parasitic response or quick-fix solution, bottled water is an established, ongoing participant in these emerging realities of water supply and access in Chennai today. The typical depiction of bottled water as a wasteful Western luxury fails to attend to the reality of water provision and access among a significant proportion of the world's population. These are situations in which bottled water has gone far beyond a status or convenience object and has become instead a normalized, even crucial means of water delivery for many people, both affluent and less so. This is not to say that bottled water is not complicit in producing the difficulties in accessing reliable, good-quality piped water that many Chennaites face—far from it. Bottled water is caught up in uneven and inequitable systems of water supply. It is a key actor in water-scarce realities. But at the same time as it enacts scarcity, bottled water also enacts abundance, offering safe, reliable, and convenient water delivery at a price. This is the complex, paradoxical face of water realities in contemporary Chennai: specific, contingent, and continually emergent.

6 Practices of Economization: Recycling Plastic Bottles in Hanoi

In this chapter we move from drinking to disposing and consider the shadow realities of bottled water markets—their inevitable "overflowings" (Callon 1998). Our explicit concern is with the afterlife of single-use water bottles in Hanoi and with the practices and calculations in which bottles become enmeshed once they have been discarded.[1] In earlier chapters we investigated how bottles reconfigured everyday water arrangements in Bangkok and Chennai, creating new drinking norms, water hierarchies, and forms of scarcity—but what of the containers that deliver this water? What of all those accumulating empty plastic bottles once the water is consumed? Their use value as packaging may be exhausted, but their brute material presence remains. Although the drinker may have thrown away the bottle and severed all relations with it, it still has material capacities and effects because the bottle is at once cheap, portable, and disposable, but also stubbornly nonbiodegradable and nondisposable. This material complexity and recalcitrance are part of the extensiveness of the event of bottled water, part of the emergent networks and realities that bottles generate as they become empty, discarded things.

The standard economic category for discarded plastic bottles is "externalities." This category assumes that the waste consequences of bottled water production and consumption are completely separate from market realities; that bottles becoming rubbish are outside the frame of market calculations. Philippe Pignarre and Isabelle Stengers (2011, 17) put it like this: "Capitalism is what never stops inventing the means to submit what it deals with to its own requirements—and the consequences don't concern it at all: it externalizes them (others can pay), or defines them as potential matter for new operations." This tension between externalizing the negative consequences of markets and turning these consequences into new market opportunities is at the heart of the relationship between disposability and plastics recycling that we explore here. Disposable plastic bottles whose exchange value is exhausted after a single use manifest the absolute banality of immediate consumption as both imaginary plenitude and rapid turnover. They also manifest the problem of material endurance: the reality of a long afterlife that is congealed in the unbreakable plastic container even before its contents are consumed. The work of marketing and framing suppresses the career of plastic packaging, its shadow reality as waste: it amplifies the qualities of the water and the bottle's disposability as convenience in action. However, the shadow reality is difficult to suppress. In markets organized around single-use consumption, waste is immanent to economic practices. While the PET bottle generates a material semiotics of expediency and impermanence—the unbearable lightness of disposable being—as soon as the water is drained from it, it emerges as a useless solid object, something that needs to be got rid of. The use of the triangular "recyclable" logo on the bottom of many water bottles is one way in which this tension is supposedly managed. This logo acknowledges the bottle's reality as rubbish and offers the consumer reassurance that that reality can be averted if consumers "do the right thing" and make the bottle available for new valuations. The recyclable logo is a device designed to qualify the negative effects of disposability.

Beyond the logo, however, is a complex reality whereby bottles are removed and managed as urban solid waste, and exactly how this happens in villages on the outskirts of Hanoi is the subject of this chapter. Although these forms of waste management are diverse, and their social and environmental effects are complex and multiple, the immediate point is that heterogeneous infrastructures make bottled water market operations possible. Disposability is successful only if all those discarded bottles are removed, if markets can be created in spaces relatively free of accumulating mountains of plastic waste and its unsettling persistence, and if the spaces for waste management are clearly demarcated from the spaces of consumption. This is not to say that these spaces of waste aren't intricately connected to spaces of consumption through complex networks of collection and removal; rather, the pragmatic demands of bottled water markets structured around disposability are to ensure that their overflowings don't threaten market frames and calculations, that accumulating packaging waste doesn't prompt reflexive interrogation of the market and its troubling consequences (Callon 1998, 18; Hawkins 2011, 535).

In the case of discarded PET water bottles, this pragmatic demand—to keep their afterlife out of market frames—has been increasingly difficult to execute. The relationship between the growth of bottled water markets and massive increases in the volume of PET bottles in waste streams has prompted numerous government reports around the world and various forms of anti–bottled water activism (for examples, see Waste & Recycling Action Programme 2006; Manly Council 2009, 2010). More troubling has been analysis of the material form of this solid waste, the results of which underscore the persistence and toxic effects of plastic pollution. There is no question that the rise of bottled water consumption has fueled escalating concerns about the global accumulation of plastic waste, about oceans swarming with plastic gyres, about plastic's endless afterlife in landfill, about failing recycling schemes.[2] Yet, while this macroperspective is useful, it can make an investigation of what happens to bottles in one Vietnamese city seem minor, even trivial. In the interest of raising ecological consciousness, the world is represented as drowning in plastic. Discarded plastic is presented as a global environmental problem, making it seem that flows of plastic waste are extensive and complex, and always greater than the sum of multiple local cultures of disposability and waste management. While there is no doubting that flows of discarded plastic are extensive and often interconnected, they are also qualitatively different and situated, and this specificity and singularity must be attended to. The mobilization of scale through invocations of the global—of global plastics accumulation or of a global environmental crisis—abstracts and homogenizes plastic waste. It reduces it to an emblematic marker of ecological catastrophe without investigating exactly how the diverse material realities of plastic bottles as waste are performed in particular places and with particular effects.

Our aim is to resist invocations of this scalar hierarchy—of global versus local or of an international plastic waste crisis—by looking at how discarded water bottles are managed in one place. Just as we investigated the specifics of water scarcity in Chennai and how bottled water has become implicated in creating this reality, here we want to understand how rapidly growing numbers of discarded plastic bottles have been incorporated into Hanoi's heterogeneous refuse-disposal and waste-recovery systems, and reconfigured them in particular ways. Although the discussion focuses on one place in detail, Hanoi's "plastic villages" on the outskirts of the city, these villages are not offered as illustrative examples or as representative of a much larger phenomenon. The plastic villages are products of the distinct history of waste management in Hanoi and must be considered a phenomenon in their own right. In other places, discarded water bottles are managed very differently, by means of processes ranging from incineration, to burial in landfills, to the imposition of container deposit schemes. This is not to deny that the recycling villages are connected to other plastic waste regimes and sites of hazardous plastic accumulation in Vietnam and elsewhere; rather, it is to stick to the details and trace how the plastic bottle as waste in Hanoi generates a complex topology of associations, practices, and values. It is to insist, as John Law (2004b) argues, that the specific and the concrete are where global complexity is located.

Hanoi's Plastic Villages

Most of Hanoi's discarded water and other plastic bottles are recycled in the so-called plastic villages located in the Thanh Tri district on the suburban outskirts of the city (DiGregorio 1994, 93). This region historically is an industrial area that has produced craft goods and other products for the city. However, over the past forty years it has become dominated by recycling industries, with different villages specializing in different waste materials. This change in the economic activities of the villages, from craft production to materials recovery and recycling, is connected to wider changes in Vietnam since the Doi Moi reforms introduced in 1986 (Mitchell 2009). These reforms prompted greater economic liberalization and saw the growth of a major manufacturing sector; they also connected Vietnam to transnational economic flows and import cultures. Rapid urbanization and growing levels of consumption were significant related developments that have had major impacts on rising levels of solid waste per capita. Apart from an increase in the amount of waste, another significant indicator of economic growth has been changes in the material form of this waste. Since 1990, the plastics industry in Vietnam has grown rapidly from a very small base, with around 55 percent of plastics currently produced in the country used for packaging (Rucknel 2006). As a result of these transformations, more and more plastic materials, from polystyrene food trays to plastic bags to PET bottles, have insinuated themselves into everyday life in Vietnam, and from there into waste streams.[3] These discarded plastic materials are part of an emerging culture of consumption and disposability in urbanized Vietnam and have become a major focus of informal waste-recovery and recycling industries in the Thanh Tri district. Hence the delineation of specific villages focused almost exclusively on dematerializing discarded plastic of all varieties, not just bottles, and making it available for new uses and values. Hence also the colloquial references to these sites as the "plastic villages."

According to a 2009 study from the Vietnam Urban Environment Company (URENCO), in the village of Triêu Khúc, the site whose activities we explore in this chapter, seventy-seven businesses and around three hundred people work in plastics recycling. The recycling businesses generate around 50 percent of the village's total income (URENCO and ENTEC 2009, 6). Different businesses in Triêu Khúc specialize in different plastics, so that the spatial distribution of plastic materials in the village manifests a complex classificatory system: specific households and businesses are identified by the type of plastic waste piled up in their yards. There are plastic bottle businesses, bags and general plastic litter businesses, hospital waste businesses, and many more (Pearse 2010).

Although we focus primarily on the highly clustered processes of plastic bottle recycling in Triêu Khúc, this village should be considered not so much a location but a sociospatial site. By this we mean that the spatiality of the village is constituted through practices and multidirectional relations. The village is both a plastics enclave and a crucial translation zone in the flows of recovered plastic waste materials moving out of Hanoi to the village, and the flows of reconstituted plastic feedstock or "new" plastic objects moving out of the village to a multitude of local, national, and transnational sites. In other words, the practices of recycling coalesce into diverse spatialities and networks. The space of the plastic villages is not "out there," or objective; it is achieved through the practices, materials, bodies, and so on that are configured there, and through the ways in which these configurations are organized and repeated. The overwhelming presence of plastic in the villages—coming and going, stored and accumulating, blowing around and burning—gives the space a powerful materialized force (see figure 6.1). Plastic seems to touch everything, and recycling practices involve managing this spatial presence, controlling the capacity of the plastic to take over, keeping it moving.

This material presence and the multiplicity of activities that surround it evoke what Theodore Schatzki calls the "teleological foundation of spatiality," or the ways in which people

Figure 6.1
Plastic everywhere.
Source: Warwick Pearse.

set up settings and locales with an eye to the activities that will be performed there and the ends people will pursue in performing these actions. In the end, teleology underlies spatiality because spatiality is the pertinence that objects around have for human activity, and the pertinence of the world around for activity ultimately rests on the matters for the sake of which people act. (Schatzki 2009, 38)

Schatzki's teleological perspective helps explain the emergence of Hanoi's plastic villages and their transformation from sites of local craft production to sites of plastic waste recovery and recycling. The historically recent shift to plastic recycling in these villages can be considered an innovation in practice (Shove, Pantzar, and Watson 2012, 85) in response to the changing "pertinence of objects," specifically the rapidly growing amounts of plastic in Hanoi's waste stream and the economic possibilities this plastic burden presents.

However, Schatzki's account of teleology is overly human-centered. It assumes that responses to plastic waste are largely instrumental and opportunistic, that plastic is an object of human entrepreneurialism—something to be acted on in the interests of economic gain. Even the most cursory observation of the practices in play in the plastic villages challenges this assumption. In these sites, plastic waste appears profoundly *im*pertinent. Although it may be a potential source of value, this value must be wrought in practice, and in these enactments discarded plastic bottles in particular become active and demanding participants. In the recycling practices evident in Triêu Khúc, the destruction of empty bottles requires significant human and mechanical effort. The bottles reveal material capacities that play a key role in shaping how things will be related and the sorts of processes that will be performed. In this way, then, teleology—or the idea that accumulating bottles cause human actions—gives way to assemblage and a focus on the ways in which human, material, technical, and other entities become associated and generate forms of distributed agency through their complex and various interactions.

Just as chapter 1 explored how the PET bottle became a market device in the beverage industry, prompting predictable and unpredictable patterns of emergent causation, bottles in the heterogeneous recycling assemblages evident in Triêu Khúc are similarly active. The "style of structuration" of these assemblages, to use Jane Bennett's (2010, 23) term, is not linear. Even though recycling practices appear to successfully coordinate processes of bottle destruction and their reconstitution into other things, other elements and disruptive emergent forces are also always at work. The intentionality of an assemblage is not the sum of its actions and effects.

Recycling as a Practice of Economization

The explicit intention of plastic bottle recycling in Triêu Khúc is to dematerialize the discarded containers to make their material components available for new manufacture. Breaking up the bottles transforms them into a reconstituted "raw" material, or plastic feedstock, that can be used in new plastics production. PET bottles have good potential value as feedstock because the original quality of the plastic is generally high—it has to be, to safely contain water. In seeking to understand recycling, we are interested in the practices and dynamics of enactment, or the ways in which recycling is performed and organized, and how different forms of enactment generate specific calculations about the material problems and possibilities of empty bottles as a potential source of new value. Another way to put this would be to say that, following Koray Çalişkan and Michel Callon (2009), we are interested in bottle recycling as a distinct "practice of economization" oriented toward the creation of value and new forms of circulation and exchange. Çalişkan and Callon argue that valuing and the organization of exchange have to be done or practiced, that value isn't so much found as created, and that there exists a multitude of ways in which this can occur. These practices of economization are embedded in shifting and highly situated assemblages—they aren't exclusive to some kind of separate realm called "the economy" (Çalişkan and Callon 2009). They are also performative in the sense that the techniques and devices used to calculate value do not so much measure an intrinsic quality but rather are actively implicated in creating it.

A focus on practices of economization also demands close attention to the stuff itself, to the material properties of discarded plastic bottles. Of course, these material properties are not fixed—they are realized in different associations and practices—but the issue with recycling is how the transformation from waste to resource is actually carried out, and the ways in which material properties enable or resist practices of value transformation: how materials might have a say in their own valuations. Related to this is the way in which specific material properties that are enacted in other contexts, such as consumption, might interfere with or persist in the context of waste recovery and recycling. How does the practice of single-use disposability that shapes consumers' encounters with bottled water interact with the practice of recycling? Obviously, this occurs through single-use bottles' contributing to the massive growth in the amounts of solid waste cities now produce. However, beyond this relation, how do practices of disposability reverberate in the ways in which recycling emerges as an economic practice and becomes meaningful?

Standard accounts of recycling usually represent it as sequential or linear—after consumption and disposal comes the work of recycling—but a close investigation of the actual recycling practices conducted in Triêu Khúc complicates this representation. The other realities of the discarded container persist and haunt new contexts. In the case of the disposable plastic bottle, they link the temporality of convenience and transience to the long duration of plastic accumulation and material persistence. In this way, recycling is recursively connected to the multiple other material practices and realities in which bottles are involved. Another way to explain this recursivity is to say that the connections and relations made present in recycling emerge not as a linear sequence but topologically. According to Mike Michael and Marsha Rosengarten (2012, 1–2), topology does not see time and space as external frameworks but as emergent: "transformations of the relations between points are not causal or linear, but open and immanent." This perspective makes it possible to understand how the spaces of plastic waste disposal are both distant—in the sense that they are militantly kept out of market frames—and also proximal, pressing in on processes of production and consumption in ways that disrupt the equation of disposability with rapid disappearance. The specific enactments of the plastic bottle as an object of destruction and recycling are folded into its other spatial and temporal enactments in markets, consumption, production, design, and more.

Another concern is to understand how recycling enacts economic subjectivities. If domestic recycling programs in the global north make environmentally aware or virtuous subjects, as underlies claims that recycling households "do the right thing," what kinds of subjectivity are constituted for those who physically collect dumped plastic bottles or transform them into pellets or plastic feedstock? What kinds of subjectivity are produced when the imperative for recycling is not moral but economic? Scavenging and recycling in Vietnam, and in many other places in the global south, are generally not understood as environmental practices but as practices of poverty or informal entrepreneurialism. While a significant body of anthropological and geographic work on waste scavengers and traders focuses on patterns of social exclusion and exploitation,[4] we want to shift the focus to investigate how an economic subjectivity is practiced. What kinds of activities and techniques are involved in the cultivation and performance of a scavenger or recycler? Rather than assume that subjects are completely subjected by the exploitative logics of economic processes, we want to understand how plastic bottles and other diverse elements are implicated in constituting economic subjectivities. This is not to deny the dynamics of poverty, profound inequality, and exploitation. Instead, it is to open up another line of analysis, one concerned with the ways in which relations among discarded plastic bottles, technical devices, spaces, and human competences do not express a pregiven economic subjectivity but rather help to realize it.[5]

Contexts of Recycling: First-Stage Plastic Bottle Recovery in Hanoi

To understand bottle recycling as a practice of economization, it is necessary to focus narrowly on actual processes. As we have argued, the value of bottles is not a result of human appraisal or intrinsic properties; it is neither subjective nor objective. Instead, it emerges from the dynamics of specific valuation practices. According to Fabian Muniesa (2012, 32), valuation should be considered "in the sense of a process, a form of mediation." For discarded plastic bottles, these processes and forms of mediation are dispersed and diverse. The actual processes of recycling in the plastic villages are connected to myriad other waste disposal and recovery practices in Hanoi, particularly the diverse formal and informal waste-collection processes that involve recovering plastics (and other recyclable materials) to sell on to the plastic villages. Although our primary concern is with what happens to bottles at Triêu Khúc, here we briefly outline the network of relations in which bottles participate before they end up in the village, to show how the dynamics of valuation shift and change along with the circulation of the discarded bottles.

Like many convenience products consumed on the move, empty PET water bottles present major challenges to urban waste administrations. Unlike domestic consumption, with its fixed systems of disposal and removal, mobile consumption requires extensive provision of public bins where packaging can be discarded in passing. Without bins, it usually becomes litter. This is why, historically, the rise of disposability, mobile consumption, and moral concerns about accumulating urban litter often have gone hand in hand. In countries where public urban spaces are highly regulated, street garbage bins emerged as one of the key mechanisms for managing this new waste reality—for keeping the public arena tidy. Bins became an expression of civic culture and a form of subtle social regulation. They suggested good behavior by functioning as a disciplining technology that prompted consumers to regulate their actions. Public bins mediated forms of citizenship by encouraging consumers to take responsibility for their packaging rubbish in the interests of ensuring moral order and "tidy towns."

This regime of bins and public infrastructure is highly situated, and particular to certain forms of political administration. As Dipesh Chakrabarty (2002, 66) notes, it is not exclusively Western; rather, it is more about cultures of modernity and the transformation of open urban spaces into public spaces, where common ideals of hygiene and aesthetic order are expressed. Keeping waste out of public spaces is one of the myriad ways in which modern citizen cultures have been built. What, then, of places where such regimes are not in place, where open urban spaces are not considered "public" or "common" in the same way? How is the detritus of disposability and mobile consumption managed? How is litter apprehended?

Hanoi is one of those places. As Lisa Drummond (2000) notes, the presence of a diverse footpath economy, often run by householders with street frontage or by rural itinerants, blurs the line between public and private. However, rather than view this different street and footpath spatiality as the subaltern other to modern, clean, citizen cultures, the challenge is to understand how this particular urban reality creates different understandings of responsibility and order, different sociomaterial relations with visible litter. While more bins are being introduced into some major Hanoi streets by urban waste authorities, along with antilitter campaigns that equate tidy streets with national economic prosperity, this kind of waste infrastructure remains a rarity.[6] Most domestic and business waste is collected by a mixed assortment of waste workers engaged in different forms of exchange relations, from municipal street cleaners to scavengers at landfills, to junk buyers paying for recovered waste materials. In rural areas outside Hanoi and other major cities, there is still extensive disposing of detritus in streets and rivers. However, within the cities waste has potential value, and it attracts diverse human interests and actions, and gives rise to vastly different cultures of disposal. The long history in Hanoi of junk buyers and waste scavengers moving through the city and making a precarious living out of recovering recyclable or reusable waste materials for sale means that many waste materials become sources of livelihood for the urban poor. This doesn't mean that these collectors do not face stigma and social opprobrium because of their intimacy with waste. Rather, the presence of scavengers and other human collectors mediates relations to urban rubbish in ways very different from chucking stuff in bins—public or private.

While collectors and scavengers represent an important element in waste recovery in Hanoi, the city also has a complex network of governmental or municipal waste infrastructures. Much of the mixed street and residential waste in Hanoi is collected by teams of municipal street cleaners, refuse collectors, and transport units employed by URENCO (see figure 6.2). It is then taken to various disposal and landfill sites. As mentioned, the other source of significant urban waste management is the "recovery system" (DiGregorio 1994, 68), comprising an assortment of junk buyers or collectors, scavengers, and receivers, all of whom use recovered materials to forge a living. All these groups are involved in distinct practices and relations with discarded bottles before the bottles reach the plastic villages for dematerialization and recycling.

Figure 6.2
URENCO street waste collector.
Source: Gay Hawkins.

This waste-recovery sector has a complex hierarchy (Mehra et al. 1996; Mitchell 2009), with collectors at the bottom. There are three types of waste collectors: dump pickers, who collect from the streets and at central dump sites and sell on to junk receivers; scavengers, who collect waste at landfills; and junk buyers, who collect from households, offices, and construction sites. Most of the dump pickers and junk buyers are women who have no other access to income. They generally work ten hours a day and walk around fifteen to twenty kilometers each day. There are currently over one thousand waste pickers at the main landfill site. They are given access to it for three to four hours, beginning at midnight, when trucks are not dumping.[7] These groups collect anything recyclable. In the case of plastic, the scavenged items include PET bottles, agricultural containers and rope, plastic bags, domestic packaging such as food containers, trays, and wrapping, plastic furniture, and a multitude of broken plastic household items.

The recovered plastic material is then sold to various waste intermediaries, or "fixed location receivers," as Carrie Mitchell (2009, 2635) terms them. These small businesses operate in a number of different sites in Hanoi. Those in the heart of the city often work on the footpaths or in backyards. Because of a lack of space, their capacity to add value to the waste is limited, and they mainly crush, package, and sell waste downstream. Larger junk shops or traders operate on the perimeters of Hanoi. Their primary function is to sort, wash, and classify the waste that is bought from the collectors or receivers and then sell it to local recyclers or export it to China. These larger sites handle all sorts of recovered materials, from plastic to paper, steel, and lead. In the case of plastics, one of their key functions is to sort them into different types: high-density polyethylene (HDPE), low-density polyethylene (LDPE), PET, and polyvinyl chloride (PVC). As elsewhere, plastics pose a significant challenge for recycling because there are so many different varieties with different chemical and physical properties, different colors, and different melting points. This material complexity and diversity means that most plastics cannot be recycled together. As Kaveri Gill (2010, 14–15) notes, "Even in advanced countries, the state of present technology dictates that plastic waste must be manually and meticulously segregated into finer polymer types before recycling, with any residual contamination resulting in an unstable final compound."

These participants in the municipal and informal waste-recovery sector play a critical role in plastics and other waste management in Hanoi.

As many studies (for example, Gill 2010; Godden 2011) have shown, they are maintaining the amenity and functionality of urban environments and are at the front line of engaging with materials as they are discarded. The labor of collecting and picking over is part of the process of enacting urban waste disposal *and* materials recovery. It also enacts an economic subjectivity for the collectors. In the interactions and interface between the discarded materials and the waste picker, a particular mode of valuation is performed. The collector is continually engaged in the "pragmatics of valuation" (Çalişkan and Callon 2009, 388), in calculations about what to collect and what might have potential for new value (see figure 6.3). It is a precarious practice, performed over and over and subject to fluctuating amounts of waste being generated, competition in wider recycling markets, and changes in urban policing and regulations. While the work involves skill and precision to quickly identify and sort materials and collect fast, the direct contact with waste at the immediate point of disposal amplifies the collector's already stigmatized identity. Collectors, like the waste, are embedded in broader social and historical regimes of classification. However, the point is not that these regimes determine individual economic disadvantage—that the individual is contained by structures. Rather, the distinct techniques and practices of valuing involved in scavenging plastic bottles and other items are situated, and this context—this specificity of local arrangements, practices, and classifications—constrains the modalities of valuation in distinct ways. As Çalişkan and Callon (2009, 384) argue, the shift from a structuralist approach to a focus on the actual pragmatics of valuation involves close attention to the conditions of complexity and mobility between and among things, people, and their context.

Although waste collecting and receiving are very distinct components in the wider practices of economizing plastic waste in Hanoi, the possibilities for enacting significant new value for the bottles in these processes are constrained. There are several reasons for this. First, as the bottles are removed from the waste stream, they are not being physically changed very much except through sorting or crushing in some instances to make them more transportable. In this way, their material market form persists and limits requalification processes: the bottles are being recovered and put into circulation, into new networks that make them available for other careers apart from waste. This material persistence also limits how many bottles individual collectors can handle and how many can be stored in receiving

Figure 6.3
Waste recovery worker.
Source: Gay Hawkins.

sites, as space in overcrowded Hanoi is very limited. Empty bottles might be light, but they are very bulky. These constraints highlight how materiality is an active participant in shaping processes of first-stage recovery and valuation, and how circulation and movement are fundamental to continuous requalification.

While it may seem that discarded plastic bottles in piles at the front of shops, lying in the street, or dumped in landfills have become waste—that is, objects that have run out of value—this classification is not fixed. The context of the bottles may signify this, but signification—like valuation—is an action and a relation. For waste scavengers, dumped bottles are things awaiting "recovery." Their apprehension of them as potential objects of valuation emerges in their interpretations of piles of rubbish, and the ways in which these interpretations or assessments simultaneously survey a waste reality and provoke a value reality. The point is that the value of the bottles is neither subjective—a human projection—nor objective; it is pragmatic. It emerges from the practical exchanges and interrelated actions that constitute waste scavenging. Muniesa (2012, 32) captures this idea of valuation as an action and as pragmatic: "this idea of 'as an action' should be understood in the sense of a process, a form of mediation, of something that happens in practice, something that is done to something else ... value is definitely not something that something just has."

Disassembly: Becoming with Plastic in the Plastics Villages

Once the bottles and other plastic waste are delivered to recycling businesses in Triêu Khúc, the aim is to materially transform them in ways that

Figure 6.4
Crushed bottles awaiting recycling.
Source: Warwick Pearse.

make them suitable for sale as plastic feedstock for local remanufacture as more plastic objects or export. Turning recovered bottles into a resource for diverse forms of plastics manufacture involves executing distinct practices of destruction or dematerialization and valuation. As mentioned earlier, specific techniques of valuation have already been engaged in the practices of waste recovery and waste trading in Hanoi. Before the plastic waste gets to the villages, it has gone through several forms of exchange as it passes

Figure 6.5
Delivering bottles to the village.
Source: Warwick Pearse.

from the disposer to the collector to the dealer, and it has undergone minor material transformations such as compacting and sorting (see figures 6.4 and 6.5).

When the plastics get to Triêu Khúc, they enter into multidirectional exchanges and undergo significant material transformation. In the plastic villages, the constitution of new value emerges primarily from the inter-actions between people and recovered plastic that are concerned with destroying any semblance of the plastic's original object form. Unlike col-lecting, recycling has to render remaindered stuff "formless in order to be reformed" (Gabrys 2011, 138). These recycling practices involve different people, spaces, actions, and calculations from those involved in recov-ery—often, relatively well-off households and other businesses that are considered at the top of the waste recovery hierarchy in terms of income and status. These practices also highlight how the complex relationship between matter and form—evident in so many plastic objects, not just the PET bottle—reverberate in recycling processes.

As we saw in chapter 1, in the production of a PET bottle, "matter and form are generated in one single gesture" (Bensaude-Vincent 2013, 20). In a microsecond, a small murky pellet of PET is stretch-blown into a PET bottle form that reveals the remarkable physical qualities of this plastic as trans-parent, light, and incredibly strong. In recycling, this relationship between matter and form presents specific challenges: getting back to the matter of the bottle demands significant interactions with its toughened form. The "raw material" is not easily accessed because it is structurally integrated into the bottle, which is designed to be indestructible.

Disassembly involves an enormous variety of sequenced actions that often require extended and intimate contact with bottles. The recycling of PET bottles entails a unique choreography that is quite different from the recycling of food packaging or discarded agricultural bags, for example. For a start, it involves working with the brute materiality of the bottle in order to destroy its object status. In recycling, previous material qualities such as the extreme durability of PET now present as physical recalcitrance that requires considerable effort to transform. In the context of supply chains and mobile consumption, the bottle needs to be unbreakable and uncrush-able, but these affordances emerge as powerful constraints in the creation of new value. Similar constraints are presented by elements of consumer product design connected to cheap production, brand differentiation, and attracting the attention of the consumer by using different-colored plastics. The bottle is designed for market realities, and this profound blindness to its afterlife is finally confronted in recycling. In this way, the bottle's physical reality as consumer packaging or a throwaway object for drinking on the go is not separate from its reality as waste; both realities are proximate and potent at the site of recycling. In the stubborn persistence of the object, the idea of disposability as manifested in rapid transience or disappearance is completely disrupted.

The first action in disassembling a PET bottle involves removing the labels and branding information that were affixed to the bottle in a microsecond by a machine; removing and sorting the caps, often into different-colored plastics; and cutting off the top of the bottle, where the different PVC plastic of the cap leaves a lip. This different plastic presents real problems for the recycling process, as it can contaminate a batch.[8] The next action is to sort the stripped, clean bottles into clear or colored PET piles. This sorting and classification into different types of plastic and different-colored PET is central to maximizing the value of the recycled material. Clear PET flakes have higher value because they can be used in the production of new PET. Colored PET flakes are more likely to be used in the production of downcycled materials, or they are melted and turned into fiber to be used in the production of carpets or polar fleece. As figures 6.6, 6.7, and 6.8 show, this work is often done by women, children, and young men sitting in backyard worksites surrounded by ordered piles of plastic bottles and lids in various stages of destruction. The practice of stripping the bottles involves a complex choreography of relations among humans, bottles, spaces, tools, machines, and competencies. It also involves constant calculations about the various types of plastic present in each bottle and the future destination and possible careers for these plastics. In households where PET is recycled, pragmatic assessments about value emerge in the way bottles are broken up and sorted; these assessments do not uncover a preexisting value but rather assemble a new one.

In the next stage, the bottles are crushed and then mechanically shredded or chipped into PET flakes. This activity is usually undertaken either in the same space as the sorting and stripping or in adjacent machine rooms, which are extremely noisy. The chipping machines are rudimentary and hazardous, but they are essential. Mechanization is a fast and efficient technique for finally destroying any semblance of bottle form (see figure 6.9).

Figure 6.6
Boy stripping off labels.
Source: Warwick Pearse.

With other plastic objects, such as discarded toys, hammers can be used to break them up; however, PET bottles simply bounce back in this encounter: they demand a machine.

Most of these steps in recycling occur in household backyards and sheds, and numerous strategies are operationalized to maintain a boundary between domestic space and plastic space. In some households, this spatial distinction is successfully established, and very clear classifications are evident, such as inside versus outside or the front versus the back of the house. In others, it is more difficult to establish where everyday life and recycling practices begin and end, for they are completely intermingled across many of the household's spaces (see figure 6.10).

Figure 6.7
Mother and child preparing bottles for shredding.
Source: Warwick Pearse.

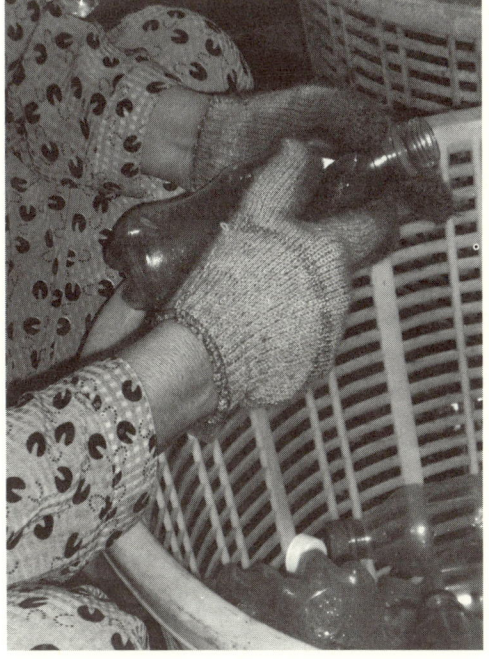

Figure 6.8
Cutting off the lip left by the lid.
Source: Gay Hawkins.

Figure 6.9
A chipping machine for reducing bottles to PET flakes.
Source: Warwick Pearse

The mechanized chipping results in PET flakes, which are packaged in large bags to be sold as plastic feedstock in new plastics production (see figure 6.11). Chipping makes the plastic mobile; it returns the object to its material origins and makes this material amenable to circulation, new object forms, and new calculations of value. The bags of plastic flakes move into numerous networks and spaces of production. Some might be sold to businesses down the road making cheap plastic objects, from bags to coat hangers, for resale back to Hanoi; others might be exported to China, depending on demand and prices.

Figure 6.10
PET flakes drying in a backyard.
Source: Warwick Pearse.

Figure 6.11
PET flakes ready for packaging and sale as plastic feedstock.
Source: Gay Hawkins.

Flakes are generally not used to make more PET bottles because it is impossible to achieve the same optical clarity that can be achieved with raw materials. Recycled PET flakes often result in an opaque or murky plastic. Although a small amount of recycled PET may be used in the manufacture of plastic bottles, most require new plastic feedstock. Murky plastic does not have the same capacity to amplify the qualities of the clear liquid product it contains, and therefore threatens the enactment of commodity value. In this way, the anticipated future and value of the recycled material are vastly different from its past. Recycling of PET as a practice of material transformation and revaluation reorders the meanings and destination of the recycled resource. This is because the properties of the original object exert numerous constraints on how it can be changed and what can then be done with its material components once it is destroyed.

In this way, materials play a role in their valuations. This is not to claim that their value is inherent but that the intrinsic properties of the material participate in the pragmatics of valuations in distinct ways; they are active and mediating forces. This claim echoes Çalişkan and Callon's argument about the relation between materiality and value:

One way of studying the mechanism through which things contribute to their own valuation is by starting with their circulation and then highlighting the role of their physical characteristics in the changes affecting them. (Çalişkan and Callon 2009, 388)

This observation is highly relevant to practices of bottle recycling. To get from dumped bottles to plastic flakes is to wrestle with the bottle as a very durable object whose potential for new value is difficult to realize. The physical and design characteristics of the bottle play potent roles in determining how the bottle is to be destroyed and in what form its recycled materials can be used. The market expectations of the PET bottle as a perfectly transparent object greatly limit the use of recycled flakes from dumped bottles. These flakes are headed for other destinations, other sites of plastic production, even as the production of new bottles devours more and more petrochemical resources. Similarly, with the actual work of bottle recycling, we can see how difficult disposability as a waste practice rather than a consumption practice actually is. What makes PET bottle recycling such a labor-intensive practice, and therefore concentrated where labor is cheap, is the bottle's resistance to dematerialization, which places great demands on the human. The power of this plastic to demand significant mechanical and human action during recycling processes is considerable.

So too is its recalcitrance, its capacity to manifest the extended temporality of extreme durability and nondisposability.

As a recycling practice, the conversion of PET bottles into plastic flakes is not the point at which final waste is averted. Breaking up bottles is a filthy and polluting process whose consequences are considered external to the pragmatic demands of valuation in play in Triêu Khúc. Although there is no question that "new" materials are being produced and new values for discarded plastic are being created, this economic reality is only one of several in evidence. For recycling in the plastic villages is also a major waste-creating practice. The plastic materials that cannot be transformed are dumped; polluted water and plastic sludge flow into local rivers; toxic fumes are released and inhaled when plastics are melted; and millions of plastic chips blow away in the process. These occupationally hazardous and environmentally polluting effects are part of what Law (2009) calls the collateral or stealth realities that shadow the intentional practices of recycling. These realities are the ones that get "done along the way ... quietly and incidentally" (13). They show that the explicit intentions of recycling are not the sum of its realities, that this practice is surrounded by a multiplicity of other realities that are far from coherent and can contest and interfere with it through the dynamics of emergent causation.

The ontological politics of recycling as a waste-creating as well as a waste-management process are difficult to discern. It would seem that the collateral or shadow realities of recycling practices in the plastic villages persist because they remain either excluded or unacknowledged in governmental regulations and in Hanoi's enforcement of labor and environmental laws. The value of plastics recycling as essential urban waste management and as a source of livelihood appears to be privileged over and above its polluting and occupationally hazardous collateral realities.

Doing Value

Thus far our analysis has focused on how discarded bottles are drafted into practices of economization that involve distinct arrangements of materials, subjects, calculations, and devices in Hanoi's plastic villages. The explicit aim of these practices is to effect new values for bottles through their material destruction and transformation, and the generation of new networks of circulation for the plastic feedstock they become. This performative approach challenges the idea that recycling represents the straightforward management or commodification of plastic waste—that bottles passively await the realization of an intrinsic value. In contrast, scrutiny of the practices in play in Hanoi's streets, landfills, and plastic villages shows that these sites and businesses operate as "coordinating devices"—that is, as distinct but related assemblages for the calculation of new values for discarded water bottles. In these calculations, plastic bottles present particular challenges. While they are active participants in enabling mobile consumption and disposability—cheap to produce, light, and tough—as waste, these market properties mediate calculations of new value in demanding and significant ways.

This material resistance means that plastic bottles are potent agents in recycling arrangements. Following Callon (1998), the plastic villages coordinate relations between humans and nonhumans (in respect to, for example, the material of the waste, technologies, spaces, and transportation), and everything involved in this arrangement becomes a *calculating agent* to different degrees. The point is that in this recycling assemblage, agency and calculation are not privileged human possessions. They are distributed across humans and objects, and the differential calculative capacity of *all* participants emerges through the process of putting things into new relations and generating new valuations. What we see in this space is how the plastic bottle prompts very distinct recycling practices, how its materiality makes specific demands on humans, how it mediates its own valuation.

We also see how, in the spaces of calculability, bottle recycling generates economic subjects. Rather than assume that the practitioners of recycling are exploited by economic practices, recycling can be considered a subjectification practice (Miller and Rose 2008)—that is, a practice involving specific competencies and modes of valuation through which people participate as economic subjects. Of course, these modes of valuing are connected to the preservation of particular interests—in this instance, the struggle to make a livelihood or to develop the entrepreneurial capacity of household businesses. They also show the crucial role that subjects with particular competencies play in enacting value. However, this does not mean that the value of things is simply a human construction or projection. Practices of economization force us to acknowledge the complex interactions between subjects and objects in enacting value. While there is no question that human actors imbue objects with significance, it is also crucial to recognize materials and the ways in which they speak in valuation practices. As Çalişkan and Callon (2009, 391) point out, "Qualification is a co-productive process whereby things and human identities are enacted as value is created."

The social life of discarded plastic water bottles is complex and heterogeneous. Exploring their postconsumption realities in one Vietnamese city has shown how they become implicated in new valuations, how they enter into the pragmatics of urban waste removal, recovery, and survival. It has also shown how the relationships between disposability, waste, and recycling need to be considered topologically rather than as a linear process of waste "management," or as an unfolding global environmental disaster. The value of accounting for PET bottle recycling topologically is that it renders market externalities or effects as processes of emergence rather than as predetermined causality. Effects do not express hidden logics or general laws; rather, they emerge in the relational enactment of the PET bottle's various capacities. As Manuel DeLanda (2011) observes, capacities are different from material properties: they are what is actualized when things are brought into relation. Capacities, then, are always relational, and they are always an event (4).

Using DeLanda's framework, we can see how the disposability of the PET bottle is a quality or capacity that emerges in its enrolment in packaging, beverage markets, consumption, recycling, and similar activities. These various assemblages are structured by distinct spaces of possibility that shape how disposability is actualized: as high turnover, as expendable, as convenience, as single use, as stubborn material recalcitrance. These assemblages are also connected in heterogeneous relations and exchanges. Things that seem distant—such as production and consumption, commodity and waste, global and local—"turn out to be far more promiscuous and can be shown to be in far closer proximity than one might initially imagine" (Michael and Rosengarten 2012, 12). In this way, the dumped PET bottle is not a mere market externality, it is a conduit of topological relations that mixes up plastic waste with plastic production and consumption. The bottle is a medium through which the multiple enactments of disposability—as drinking practice, as gesture of discarding, as massive pollution burden, as source of precarious survival, as shifting valuation practices—become copresent and connect. It is something that shows how the ever-growing flow of plastic moves in multiple directions. As chapter 1 showed, the single-use PET bottle was designed to be wasted. In the plastic villages of Hanoi, this market qualification emerges as a major constraint in the enactment of recycling. The colored lids in different plastic, the lip that remains after the lid is removed, the durability of PET, its translucency—all these elements resist dematerialization. In this way, recycling reveals the troubling paradox of disposability, the ways in which easy come is not easy go.

III   Ethical Drinking

7 Contesting Bottled Water: Markets and Material Publics

The New Smoking

In 2008, Giles Coren, food critic for the *Times* of London, concluded an article about bottled water with the derogation, "Drinkers of bottled water are the new smokers." This scathing assessment came after a long rant about the markups on the water, the millions of food miles the bottles and water had traveled, and the growing role of plastic waste in global pollution (Coren 2008). This review was picked up by the environmental website Treehugger.com, which dumped another load of moral outrage on the hapless bottled water drinker: "The vanity of it! While half the world dies of thirst or puts up with water you wouldn't piss in" (Alter 2007).

These examples of popular vilification are not random or isolated. Over the last ten years, a phenomenal assortment of online campaigns, NGO reports, newspaper articles, YouTube videos, public art events, memes, and more have emerged attacking bottled water.[1] This activism has made bottled water controversial. There is now no question that diverse environmental publics stalk this commodity, making trouble for its markets and generating various forms of critique about the social and political effects of drinking water from single-serve bottles. Despite the extent of this contestation, however, it is by no means universal or inevitable: bottles are not under attack everywhere. As our discussion of Chennai and Bangkok showed, this mode of delivering drinking water is becoming increasingly normalized—even celebrated—in places where a safe water supply is not guaranteed. Controversies around bottled water, then, are very situated. Despite activist claims that bottled water markets threaten the universal human right to water, the bottle's reputation as a detested commodity is highly contingent.

WE'RE DROWNING IN PLASTIC.
Drink tapwater. Filtered fresh. www.toronto.ca/water | ╓╖ TORONTO

Figure 7.1
Anti-bottle activism, Toronto, 2008.
Source: City of Toronto.

Most of the opposition to bottled water is confined to places where state provision of water is both expected and emblematic of citizenship. In many of these places, it seems that no sooner had the plastic bottle of water appeared as a mass commodity as opposed to a boutique item than it became a "matter of concern" (Latour 2005). The question of what actually comes to matter in these campaigns is important because bottled water has provoked an enormous variety of issues and public anxieties beyond its purported threat to the tap. Although many campaigns focus on bottles' role in undermining public water supplies, this is often simply the starting point for an escalating set of concerns about anything from the environmental burden of petrochemical cultures to the privatization of aquifers. This diversity is not just a testament to the extent or seriousness of the bottled water problem; it also reveals the experimental nature of politics—the ways in which the same object can prompt heterogeneous concerns and attract multiple publics according to its different ontological realities as commodity, waste, or something else. What happens to bottled water markets, then, when they become the target of political activism? How did the bottle of water acquire such a bad reputation and become identified as an issue? What are the effects on consumers of making bottled water into a political object? This chapter pursues these questions.

The emergence of a multiplicity of concerns and issues around bottled water and of various publics opposed to it is another expression of the advent of bottled water. As we have argued, an event is more than a historical occasion: it is a process of becoming and effectivity that signals the proliferation of new relations, problematizations, and identities. Event thinking is nonreductionist and foregrounds the performativity of social processes, the ways in which particular actions, things, and associations can enact changed realities. How, then, did protests and vilification situate bottles in new relations that prompted queries about the interactions among water, beverage companies, and drinkers? Our aim is to document these processes—to track the bottle's coming into being as not only a political object but also a political event. By empirically tracking how bottled water was made into an issue, we can better discern how it has been transformed by political processes, and how these transformations have unfolded into broader questions concerning the ethics of drinking water. This is not simply a matter of the moral reframing of bottles from good to bad; it is about changing the relations between subjects and objects, about bringing different kinds of practices and worlds into being, so that the bottle appears as a very troubling thing. The emergence of a changed reality for bottled water shows how it is far from a unified or fixed object.

Tracking the event of bottled water as an issue also makes it possible to understand how the formation and spread of issues and controversies occur in relation to markets. Rather than seeing political processes as external to autonomous markets, our approach seeks to understand how they become immanent to them—how some markets can become embroiled in reflexive activities that question their functioning, effects, and organization. For Callon, Méadel, and Rabeharisoa (2002), these reflexive processes of issue formation and controversy reveal the contingency and fragility of markets, and their constant vulnerability to contestation. Despite the popular assumption that markets are invincible and omnipresent, dominating all areas of social life, they are subject to any number of destabilizing forces. The work of framing is one of the key ways in which markets seek to contain these forces: the frame establishes the boundaries of the market. Market frames, however, create their own structural instability: "In the sense that it structures an exterior to itself, a framing is its own inescapable source of the threat of overflows" (Çalişkan and Callon 2009, 8). In bottled water activism, these overflows have been the focus of extensive publicity and exposé in a wide variety of forums. Overflows, then, can bite back through any number of processes, from environmental impacts that become impossible to ignore to publicization and activism by concerned groups.

However, while beverage markets and corporations may be the specific target of most anti–bottled water campaigns, many such campaigns also appeal to universal values connected to water, social justice, and human rights. One of the favored wider contexts invoked in activism is an idealized or imagined reality in which water is publicly and safely available to all, and is not a private singularized resource or market thing. A crucial consideration in any analysis of bottled water activism, then, is how the specificity of the issue is situated in relation to general appeals to the common good. It is difficult to comprehend bottled water's issue networks without investigating how the bottle is positioned in relation to the broader biopolitics and governance of water: how bottles are seen to interact and interfere with more universal concerns related to the support of life.

As our investigations in part II showed, bottled water practices produce numerous ontological realities and points of interference with the particular biopolitical regimes for water provision in which their markets are located. However, the exact ways in which bottled water markets become implicated in these regimes cannot be known in advance. Bottles are not always already biopolitical; they become so according to the specific situations and arrangements in which they find themselves and the ways in which they acquire the capacity to affect more equitable water arrangements. In the settings explored in part II, political effects were surreptitious—they were part of the shadow realities of bottled water. Lurking behind the celebration of a new market in water, or the proliferation of brands offering a solution to scarcity and an abundant supply, was a shifting ontological reality in which commodities and businesses were starting to become equated with safer or more convenient forms of drinking water provision—in which bottles were becoming normalized.

In contrast to politics as a shadow reality, the explicit aim of the activism and issue networks examined in this chapter is to make present the unexamined effects of growing markets in bottled water. These networks subject the bottle and its contents to scrutiny and interrogation, and amplify the bottle's relationships with the wider biopolitics of water. Their strategic intention is to disrupt dominant market framings. What an investigation of anti–bottled water activism offers is the ability to see how more capacious democratic ideals such as the "right to water" or the "water commons" become meaningful through the process of making the bottle into a political object. Beatriz Da Costa and Kavita Philip (2008) call this "tactical biopolitics," a concept that highlights how the potency of a specific technoscientific issue, such as the rapid rise of markets in bottled water, can be used to make manifest universal concerns about the common good. The question is, how does a particular anti–bottled water campaign make present the role of water as a basic material of life? How do specific issues become implicated in the composition of the collective or the imagination of common worlds? In what ways does activism render "living" subject to ethical interrogation and query? As part II showed, the ways in which bottles make trouble for the provision of safe public water or environmental protection are immensely variable, and often unacknowledged. In these settings, PET bottles become participants in assembling new relations between water and populations that do not automatically overrule other water regimes but interact with them in complex ways. The challenge for bottled water activism is to show how interaction becomes interference: to show how bottled water markets disrupt other, more equitable and sustainable forms of water provision—how commodity form becomes destructive biopolitical capacity.

How, then, can we understand the minoritarian issue politics of bottled water as a form of tactical biopolitics? The only way to answer this question is to look at some specific campaigns and activism. Among the plethora of available examples, we chose three for the very different techniques they use to contest bottled water markets and interrogate water provision. Their differences highlight the performativity, or "arts," of the political (Amin and

Thrift 2013)—the diversity of processes whereby bottled water can be made politically calculable. In the first case, we examine the Inside the Bottle campaign of the Polaris Institute, a Canadian NGO. This campaign provided one of the earliest articulations of bottled water as an issue, and it has been enormously influential. Its website remains one of the most hyperlinked sites in anti–bottled water campaigns to date. The focus in this example is on the initial formatting or framing of bottled water as an issue, and the role of networks in publicizing it and assembling concerned publics.

In the second example, we investigate how the international water filter company Brita developed an award-winning advertising campaign by exploiting anti–bottled water activism. In Brita's 2008 FilterForGood marketing strategy, consumers were invited to ditch the bottle and switch to more sustainable filters. Strange as it may seem to consider advertising as activism, there is no question that this campaign publicized the problem of single-use plastic bottles very powerfully. What it highlights is the role of affect and vital materiality in "making things public" (Latour 2005), and the difficulty of establishing clear distinctions between consumers and publics when advertising explicitly appeals to both these identities at once. Finally, we consider Do Something, also known as the Bottled Water Alliance, an Australian-based organization that works with business and government to reduce the "menace of bottled water" (Do Something 2008). In this campaign, the reintroduction of free water fountains in public space challenged the practice of buying and carrying a personal supply. It promoted a revaluation of the water fountain as a "public thing," a democratic object shared by all.

Each of these campaigns implicated a wide range of people as "communities of the affected" (Marres 2012). They all show how bottled water is not simply the passive object of political deliberations and activist critique but also a political object—something that can acquire active "powers of engagement." This is Noortje Marres's term, and she uses it to describe forms of contemporary political participation that have moved beyond the dynamics of simply informing citizens about troubling issues to the active implementation of changes in the material practices of daily life. In these examples, engagement is both embodied and more-than-human. It requires specific objects, technologies, and practices in order to enact political participation and make publics. The challenge, then, is to empirically track the actual processes of the emergence of bottled water as both a controversial issue and a political object; to document how specific associations between and among material things, knowledge claims, concerned groups, and technologies are generated; and to examine how these associations have both problematized the bottle and given it the capacity to call publics into being. As the bottle of water emerges as controversial and problematic, it becomes capable of provoking drinkers: of suggesting anything from boycotts and practices of refusal to switching to filters, to using the water fountain rather than carrying a bottle. In this process, the bottle becomes something with the lively capacity to engage drinkers as publics and ethical drinkers rather than as consumers. It becomes not just a political problem but a political material.

The other common element in these campaigns is that they reveal the increased prominence of NGOs and networked communication technologies in new forms of politics. Bottled water could not have become an issue, and its markets could not have become so significantly contested, without the exponential growth of information and communication technologies (ICTs) and new modes of distributed activism. These networked technologies, and the platforms they have created for the staging and dissemination of political issues, are evidence of major transformations in forms of rule and political processes. They lead to new forms of power and influence that are reshaping democracy, political participation, and governing beyond representative politics. In the terrain between state, economy, and civic society, a variety of intermediating groups and practices have emerged that are "reformatting" politics (Dean, Anderson, and Lovnik 2006). Bottled water has been picked up by many of these groups with surprising enthusiasm—it is obviously a market and a commodity that troubles a number of sacrosanct principles. In making this observation, however, we reiterate that our focus here is on *how* activism is enacted, not simply why. Our aim is not so much to assess the legitimacy of anti–bottled critique and activism but to understand how they work practically—to investigate exactly how they have made bottled water calculable as a political problem and the rapid rise of a new beverage market into a political event.

Formatting the Issue: The Polaris Institute and the Inside the Bottle Campaign

Bottled water as an issue has a relatively recent history. Although markets have grown phenomenally over the past twenty-five years, opposition to them was not immediate or automatic. In 2004 there were very few articles in Europe or the United States on the growth of this beverage market and its implications. By 2009, however, the situation had completely changed (Clarke 2010). Bottled water had become controversial, with a mounting number of activist campaigns and mass media reports on the activities of the industry and the madness of this new drinking habit. This eruption of public scrutiny and controversy wrenched the product out of its cozy location in beverage markets and threw it into the turbulent zone of what Callon and colleagues term "hybrid forums": spaces where diverse groups prompt questions about issues they consider themselves affected by (Callon, Lascoumes, and Barthe 2009, 18). Recognizing the eruption of a hybrid forum, however, is not the same as understanding the complex procedures whereby such forums coalesce into an issue, become populated, and generate effects. Developing such an understanding entails mapping the various social and technical processes of issue formation and the ways in which those affected emerge as a public.

The Polaris Institute's Inside the Bottle campaign offers a rich example for understanding the dynamics of issue formatting and emergence. This NGO has played a crucial role in establishing the durable frameworks through which bottled water has come to be understood as a political problem. However, in examining the history and practices of the campaign, the danger is that it could be overdetermined as the singular origin or cause of the issue—that the complexities of an event could be reduced to the work of an NGO. While there is no doubt that NGOs have become key players in issue definition and formation (Dean, Anderson, and Lovnik 2006), their practices of advocacy must be empirically documented to show exactly how they define issues, what practical devices they use to make issues relevant, and how they call publics into being (Marres and Rogers 2005). After all, as Marres and Rogers point out, there is no issue without a public. Such an empirical approach shifts attention from an instrumentalist account of NGOs to the fundamentally performative dimensions of their work—the particular ways in which they enact politics. It situates the NGO as a powerful actor in the assemblage of controversies and publics rather than as the representative of a pregiven general political will. The NGO doesn't necessarily give voice to a public; rather, it seeks to publicize issues and generate attachment to them. In other words, NGOs make issues and publics in very specific ways. According to Dean and colleagues (2006, xxii), these groups "do not represent constituencies, although they may mobilize them.... To an extent, representation per se has no meaning; instead, interactions performatively produce and reproduce a morphing set of expectations for participants."

The Polaris Institute was established in Ottawa in 1997 after a decade of social movement activism around the two free trade agreements that fundamentally restructured the Canadian economy. As the account of their origins on the Polaris website explains, "The pivotal lesson that emerged from this social movement experience was that transnational corporations had effectively secured control over the reins of public policy making in this country [and elsewhere] to the point where citizens were becoming politically disenfranchised" (Polaris Institute 2014).

Funded by donations and employing a small staff of researchers and organizers, the organization adopted the goal of initiating and coordinating "citizen actions against corporate driven globalization." Several issues were targeted in the early years, mostly prompted by emerging controversies and political conflicts that were getting some coverage in the mainstream media. One of the most prominent was the increasing privatization or corporate takeover of public resources and services. Throughout the 1990s, Tony Clarke, founder and executive director of Polaris at the time, had been tracking this issue, especially as it played out in relation to water. As he says, "I became interested in bottled water in the 1990s as the industry began to take off. It seemed to be a vivid example of some of the larger issues around social justice.... There were only a few articles here and there, but no comprehensive analysis of the industry so we followed the money" (Clarke 2010).

Clarke's interest initially involved tracking both the rise of the bottled water industry and the emergence of water provision and security as a growing global concern. In 1996 the World Water Council was established by a consortium of government and commercial stakeholders involved in water supply. Its stated aim was to promote "awareness, build political commitment and trigger action on critical water issues at all levels," especially in ways that ensured conservation, protection and access to water in environmentally sustainable ways for the "benefit of all life on earth" (World Water Council 1996).

The members of this council were primarily drawn from government and interested commercial entities, such as Coca-Cola, the World Bank,

Suez (the French water company), and a scattering of national and local government representatives. No community or activist organizations were represented. The World Water Council organized the first World Water Forum in 1997, and then another in The Hague in 2000. Clarke attended The Hague forum, which had around five thousand delegates, and networked with the small number of people from civil society and activist groups present—only about thirty-five of them. They were also concerned about the growing role of big corporations, including beverage companies, in water provision. As Clarke says:

We began to see that this picture that we were talking about regarding the privatization of water services was not really complete without looking at the role the bottled water industry was playing. And from there, as we got into it, we discovered just how large the industry was ... and that it was the fastest growing segment of the global water industry. (Clarke 2010)

This account shows how Clarke connected the rise of bottled water to the growing controversies over water privatization and scarcity that had escalated during the late 1990s. In the early 2000s, the now famous "Cochabamba Water Wars" erupted in Bolivia in response to a failed water privatization program that had led to huge increases in prices for the poor and an unreliable water supply. Other battles over water privatization also occurred in Manila, Jakarta, and Johannesburg, and protests against the leaching of groundwater by Coca-Cola and PepsiCo occurred in India and Ghana.[2] As an NGO, Polaris both responded to a flowering of water issues and framed bottles as a particular manifestation of these larger-scale politics.

In 2007, Clarke published *Inside the Bottle*. This book was the first comprehensive attempt to critically analyze the social, political, and environmental impacts of the bottled water industry in Canada and the United States. It signaled a major new campaign for the Polaris Institute. A parallel website was launched around the same time, where people could sign up for regular *Inside the Bottle* updates; the website functioned as a clearinghouse for articles on the state of the bottled water industry and related water campaigns. The book *Inside the Bottle* was the result of extensive research into the operations of the bottled water industry that Polaris conducted alongside its wider involvement in other water campaigns, such as Blue Gold (Barlow and Clarke 2002) and the World Water Forums. Based on scrupulous investigation of the big four companies that dominated the industry and their practices, this book was a mixture of detailed corporate and market analysis, industry exposé, and political advocacy. The stated aim of the campaign was to "create awareness around key issues by conducting in-depth research and sharing information with grass roots organizations" (Clarke 2010).

This brief prehistory of the Inside the Bottle campaign is an example of the initial process of articulating and formatting an issue. The questions it raises concern what organizational and technical devices were used to frame bottled water as controversial. How were people affected by this issue identified and addressed? And how did the Polaris Institute establish its authority to speak for those opposed to bottled water? As an NGO, the Polaris Institute felt authorized to participate in emerging international forums around water. This authority was established in a variety of ways. If the expectation of activist organizations is that they be seen to act, then participating in controversies and becoming actors in them is a crucial realization of this authority. Legitimating this activity was a mixture of technical expertise in the form of detailed research and information gathering, appeals to universal values such as water rights, and critiques of large-scale abstract forces such as privatization. Also important were the history and reputation of the organization and its established credibility as a nongovernmental political participant keen to hold the state and corporations accountable (Feher 2007). During the late 1990s the Polaris Institute became involved in global water debates and hybrid forums as a participant. It established alliances with other organizations, developed specialized expertise on the bottled water industry, and put bottles on various agendas. It extended the questions surrounding water privatization to water as a single-serve commodity. In this way, the inventory of global water controversies was spread beyond pipes or the lack of them, or corporate takeovers of public utilities, to significant changes in beverage markets. The work of the Polaris Institute during this period made the rapid growth of bottled water markets emblematic of larger water politics. The particularity of the bottle was framed as another manifestation of complicity between corporations and inadequate governance in establishing or maintaining water security.

Another key way in which the Polaris Institute's legitimacy to contest bottled water was enacted was through the invocation of that well-worn political rhetorical figure, "the people." As figure 7.2 shows, the slogan for the Inside the Bottle campaign was "The people's campaign on the bottled water industry." This strategic appeal to "the people" implicitly legitimated the

NGO as the agent or representative of a preexisting popular will. However, in 2005 publics were yet to coalesce around the issue of bottled water in any significant way: they were strategically invoked in the campaign but not yet affected or implicated. In problematizing bottled water markets and advocating action against them, the Polaris Institute was not reflecting public interest but trying to generate it. "The people" was an idealized democratic collective that had yet to materialize as a collection of actual participants in the issue network. This reflects one of the fundamental structural challenges facing NGOs and their particular location in political processes: how to make the issue relevant to publics—how to populate the issue and give substance to the claim of being democratically delegated to speak for "the people."[3]

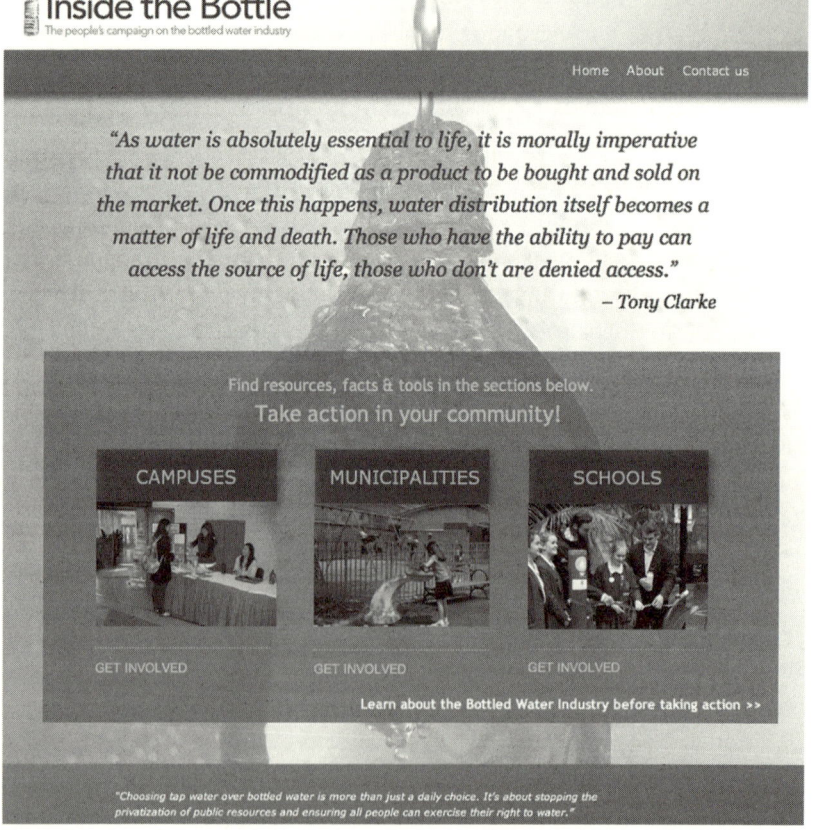

Figure 7.2
The Polaris Institute's Inside the Bottle campaign website.
Source: Polaris Institute.

To make this observation is not to endorse the common claim that NGOs are nonrepresentative, and therefore illegitimate actors in politics—"unelected busybodies" (Dean, Anderson, and Lovnik 2006, xxiv) stirring up trouble for its own sake. Rather, it shows how this form of political practice depends on arousing or sparking publics into being and generating political capacities for them. In this sense, it embodies a postrepresentative and performative notion of politics. Rather than see an issue as an expression of preexisting interests or stakeholders, a performative approach focuses on the dynamics of what Whatmore and Landström (2011, 584) describe as a "swarm." In the process of being affected by an issue, of responding to the concerns it generates, people gather around the issue and become associated as part of an emergent public.

However, in acknowledging the active role of NGOs in making things public and making publics, it is crucial to understand how their institutional and practical techniques of activism have an effect on the shape of the publics that emerge. The Polaris Institute approached its potential publics through a logic of empowerment. By researching and spreading empirical information about the bottled water industry in multiple forms—books, the Internet, listservs—the institute assumed that publics would assemble and participate in contesting bottled water markets once they understood their negative effects. This strategy of distributing information meant that the subjects invited to gather around the issue were not involved in defining it; they may have felt affected by it, but their knowledge and perspectives were not activated in framing the problem. In many senses, publicizing the issue had to function much like other forms of promotion and marketing. It had to find a mode of address and a voice that would resonate with or appeal to as many subjects as possible willing to identify with it. It had to provoke a critical imagination in potential publics about what bottles were doing to water. The issue had to be marketed in ways not unlike the marketization of the commodity: the problems of bottled water had to be made meaningful and relevant to the everyday lives of drinkers.

How, then, did the Polaris Institute's Inside the Bottle campaign reformat water politics? How did it frame the bottle as a political threat to the broader issues of water security and public provision? And in what ways did these framings generate an issue network? As mentioned, Clarke had already connected the rapid growth in bottled water markets with wider global trends toward water privatization. In presenting bottled water as controversial, this link was elaborated using the specter of the corporation. Just as large multinational engineering companies were moving into water provision, so too were the big beverage brands. In water infrastructure, global companies such as Suez and Veolia were emerging as major players; with bottled water, it was the "big four": Coca-Cola, PepsiCo, Danone, and Nestlé. In the opening chapter of *Inside the Bottle*, the history of these corporations' expansion into bottled water markets was mapped in meticulous detail, showing how water had emerged as a major new growth area for them all. Tracking patterns of industry takeover and consolidation, Clarke showed how the growth of bottled water markets was opportunistically connected to wider debates about public water risks and quality, water scarcity, and changing consumer health practices. This was an interesting connection to make, and not immediately obvious. If you believed the corporate spin from beverage companies, bottled water had nothing to do with mainstream forms of public water supply; it was merely a specific market response to new consumer demands. Water as a rapidly growing segment of beverage markets was completely distinct from taps and their essential role in sustaining populations.

Inside the Bottle disrupted this market framing by showing how beverage companies consistently positioned bottled water in relation to tap water. Corporate marketing did this in numerous ways, from making claims about the water being much tastier than "just tap water" to providing detailed explanations on labels about the unique source or treatment processes that rendered bottled water "pure." In the process of "turning water into water" (Clarke 2007, 54), beverage companies inevitably qualified their product as superior to other available forms. And, as we saw in chapter 2, brands were a crucial part of this process of building a distinct economy of qualities. This qualification process was also deeply related to another key market device: pricing. Without these elaborate justifications for the enhanced quality of water in bottles, it was hard to convince people to pay a lot more for something they were already getting relatively cheaply. Clarke (2010) put it like this:

We found that the work against privatization of water services was incomplete around the bottled water industry.... Once people get into the habit of paying significant amounts of money every day for bottled water then you start to build a culture for people paying for water no matter where it comes from. It was the flip side of privatization. In political cultures where there is a fair degree of public or government support for the tap water system, the way to undermine that is to create a counter culture to that by grossly increased bottled water sales.

Just as price functions as a key device in assembling markets, the same principle applies in the framing of bottled water as an issue. Recognizing that consumers were often "price-sensitive," to use the industry jargon, *Inside the Bottle* presented their consumption of water from bottles as an expensive con. Price was used to engage consumers, to make the issue relevant to them, and to invite reflection on the ways in which the corporation was marking up a liquid to which they already had easy and cheap access. However, price gouging was not simply about ridiculous markups, it was also about the way in which purchasing water in a bottle *re*qualified it from public resource to private good. In other words, this qualification had serious shadow realities that changed the political qualities of water. In casting bottled water as an issue, the Polaris Institute sought to trouble the consumer's relationship with the market and the state. Paying for water in a bottle didn't just mean you were being ripped off; it also meant you were replacing a relationship with the state with a relationship with a corporation, and often getting the same water. In this way the campaign appealed to consumers to resist the intrusion of the market into a sphere of their daily life where government was functioning effectively. It invited them to act as citizens rather than consumers in water choices.

The second framing that was used to problematize bottled water drew in plastic waste and environmental concerns. The unsustainability of a single-use throwaway container exposed the externalities of the market: its exploitation of nonrenewable resources, its contribution to greenhouse emissions, and its role in escalating amounts of unrecyclable plastic. Here the bottle as a material thing with a significant afterlife was made present. While the industry framed the bottle as a neutral container—transparent, convenient, and recyclable—Inside the Bottle framed it as a major problem. This framing resonated with those who were already engaged with environmental issues. In seeking to populate the issue, this angle connected bottled water to other political issues and those attached to them. Bottled water was represented as another manifestation of environmental exploitation and growing controversies around global plastic accumulation. According to Clarke, this approach was explicitly aimed at young, "environmentally aware" people who had already learned to be affected by images and information about plastic waste and pollution.

Other framings related to corporate water takings and again sought to reveal the externalities of market practices: theft of water and depletion of poorly regulated aquifers highlighted concerns about who owned water and how it should be managed and distributed. The issue of privatization was central to this particular problematization. Bottled water markets were seen as laying the essential groundwork for a more concerted attempt to wrest public control of water from state authorities and transfer it to corporations:

The Big-Three water service corporations are getting a helping hand from the bottled water industry in their quest to privatize water services in North America. After all, the industries' main competitors continue to be municipal water utilities—and much of the marketing and advertising for bottled water is designed to wean people off tap water by undermining their confidence in public utilities. (Clarke 2007, 106)

These framings show how essential the work of agenda setting is to making an issue. In the case of bottled water, the emergence of the issue was enabled by publicizing and defining, in very specific ways, the externalities of the market, from price to waste to corporate takeovers of water sources. But the function of activist frames is not just to expose and critique: their explicit political aim is to invite attachment. The framings of *Inside the Bottle* were essential to assembling diverse publics that felt affected by the issue in different ways. Just as the bottle was a complex material thing, so too was its potential to disturb. Information about its many realities outside the market provoked specific registers of concern and political calculations that had the capacity to resonate with multiple constituencies. This effect highlights how issue framing by NGOs is often shaped by the need to extend the issue's associations to as many potential publics as possible. Just as market framings seek to address multiple consumers by positioning the commodity in ways that resonate with their existing interests and identities, issue framings could be said to operate according to the same structural logic. The difference is that in issue framings, the calculative logic is political rather than economic. One area in which there may be a common calculation across markets and publics has to do with identity—though the content of these calculations is often vastly different. In inviting people to identify with bottled water as destructive, the Polaris campaign offered drinkers alternative ethical identities to those promoted by markets. Whereas markets framed drinking from bottles as a marker of health awareness and self-care, activism framed *not* drinking from them as a marker of ethical concern for the environment. Similarly, brands too became available to new calculations as the issue unfolded: brand recognition can be as central to issue formation as it is to market development. Polaris's focus on Coke, Pepsi, Nestlé, and Danone meant that those who recognized these brands as consumers were potentially open to other forms of engagement simply because of their already present capacity to be affected by them. As Robert Foster (2008, 175) argues, brands, like goods, are open to constant requalification and repositioning, and political contestation shows how brand power can be exploited in diverse ways.

Another element in the Polaris Institute's framing of bottled water was the invocation of well-established left critiques: of multinational corporations and their practices, environmental destruction, and the privatization of or decline in public services. These framings, however, were not empty appeals to abstract political scare words. What made them engaging was a meticulous empirical analysis of how bottled water markets enacted very specific forms of water privatization, environmental damage, and changes in drinking habits. In this way, familiar political abstractions were made meaningful by real examples, and by the presentation of these examples in targeted and accessible formats. "Awareness raising" was the explicit aim of this strategy. But the Inside the Bottle campaign implemented this strategy in such a way that the information did more than critique and expose: it was also *networked,* so that its political capacities and effects were amplified through the dynamics of distribution.

By using networking to publicize the issue, Polaris was able to multiply the spaces and possibilities for contestation and public attachment. Information doesn't simply inform or expose on its own; it accumulates force and political capacities according to how it moves. As an NGO, the Polaris Institute played a key role in assembling, interconnecting, and formatting the issue, but this was only the opening stage of emergence. The use of online networks hosted by Polaris was fundamental to giving the issue life. These online strategies had diverse functions that developed over time; they ranged from acting as a clearinghouse for information on bottled water to the creation of listservs that sent out constant issue updates, the deployment of e-petition and e-letter capacities to prompt user action and engagement offline, the construction of massive databases of active constituents, and the mobilization of social media to promote everything from "bottle-free days" to the online video, *The Story of Bottled Water* (Story of

Stuff Project 2010). These different forms of networking helped build the scale of the issue and generate more hybrid forums around it and more affected publics. Networking technologies allowed for highly tailored strategies, such as specific networks of students in seventy-seven universities and colleges across Canada that were hosted and coordinated through the official Inside the Bottle website, as well as connecting with social media networks such as Twitter and Facebook, which also offered significant amplification of the issue beyond targeted groups. Twitter was especially valuable, as it was followed by a large number of mainstream media, so that it provided a very effective tool for mass media outreach and also for legitimizing Polaris as an important voice for the issue. As Michael Dieter (2009) argues, the network is the medium that makes an issue legible, that contributes to stable definitions of actualized knowledge. Networked modes of assembly privilege movement and allow the emergence of distributed publics and multiple scales of action, from the small and local to the global. In this way, networking technologies and practices became crucial devices in contesting bottled water markets—in enacting issue politics.

However, public engagement could not be restricted to online environments: offline networks were equally if not more critical. As Joe Cressy, the manager of web strategy for Inside the Bottle, explained:

The overall key strategy was to maximize the number of online constituents and to engage them proactively to ensure they become advocates in their own right.... Our online data component was designed to bring people in but then the follow up piece focused on the best way to empower these online constituents so that they were not just recipients of information but actors in their own way. There had to be action.... The success of the campaign was based on individuals and individual action. It's necessary for us not just to educate but to engage and empower. (Cressy 2010)

This explanation highlights the ways in which Inside the Bottle issue networks mobilized diverse actors: how people who felt addressed by the issue or involved in it were then prompted to participate in particular activities through a variety of other techniques and devices. Once again, the NGO played a key coordinating role, but the development of effective anti-market actions was not straightforward. Through trial and error, the Polaris Institute eventually came to focus on lobbying different levels of government, prompting them to start banning bottled water in their various jurisdictions. By 2011, more than seventy-five cities in Canada had taken action by phasing out the sale and use of bottled water in government buildings. This strategy has also been adopted in many other countries. The key justification for action in this context was that governments that are responsible to populations for the safe provision of water should not also be in the business of promoting the private sale of water. Similar strategies were developed in universities, with several campus student bodies working with administration to create "bottled water–free zones" by removing bottled water for sale. A similar campaign in high schools then followed, targeting the teachers' union and lobbying it to challenge the sale of bottled water in schools.

These actions can be described as forms of boycotting designed to contest the operations of bottled water markets by morally shaming the public institutions that supported them. Government bodies, universities, and schools were effectively implicated in this way because they already had a mandate to serve the public and protect the public's interests. Selling bottled water in these organizations was seen as undermining their duty to support this public interest, and also the principle of water as a public good. Equally significant was the way in which Polaris targeted public institutions rather than the producers of the product, the beverage corporations. After all, many of them—but most of all Coca-Cola—were already under siege from an enormous variety of issue networks targeting everything from the treatment of workers in Colombian bottling plants to spurious PR campaigns around HIV AIDS in Africa.[4] In lobbying those who hosted or implicitly supported bottled water markets, the issue networks shifted their focus to undermining cultures of privatization—to challenging the normalization of water being delivered in bottles when taps were readily available.

Another critical effect of these actions was that they provoked more publicity for the issue. Banning the bottle from government institutions attracted lots of mass media attention. As Clarke explains:

Once you get a municipal council or city government exploring some regulatory or legislative action it raises it to a new level. You can get a fair amount of public interest going in a given jurisdiction around this. If it becomes a story that can spread across the country then that is great ... that was certainly the case with the city of Toronto and their action in banning bottled water it became national news. Then Toronto took their legislation to the Federation of Canadian Municipalities and were able to get it on the agenda and it was passed.... So you can build momentum and take advantage of opportunities. (Clarke 2010)

This description of the momentum of an issue is compelling, but it is somewhat linear. What Clarke elides is the ways in which issue networks depend on movement, contingency, and change, *in all directions*. It is not so much a question of the issue snowballing but of it becoming a networked assembly involving complex dynamics and devices, working in many different registers, that keep the issue in an ongoing state of emergence. The mass media amplified the Toronto decision by reporting it, and this public attention enhanced this local municipality's authority to lobby the National Federation. At the same time, these media actions also—inevitably—attached new readers to the issue, who then visited the Inside the Bottle website. The contours of controversies are topological. What the Polaris Institute's campaign shows is that political action is not simply about tactically opposing a fixed object of contestation; it is about the ongoing assembling and networking of the issue in ways that keep it legible and alive, and that maintain its force as an event. As Dieter (2009, 58) argues, the rise of issue networks demonstrates "how emergent, material, collaborative and complex dynamics have become important topics for thinking of new ways of acting politically."

This account has documented the crucial role of framing and networked assemblages in the emergence of bottled water as an issue. As an NGO, the Polaris Institute connected the rise of the bottle to broader developments in global water politics. Its work in representing growing markets in single-serve bottled water as a particular manifestation of water privatization had the effect of requalifying the commodity and turning it into a matter of concern. This work of framing and mediating the issue shifted attention from the bottle's reality in markets to its multiple realities beyond this very restricted field. However, making this issue of concern matter beyond the confines of an activist NGO required a range of technical devices and publicization processes focused on attaching publics to the issue and prompting them to act. Through the dynamics of research, the provision of counter-information, networking, and the like, the bottled water issue was populated by inchoate and distributed publics that felt implicated in its threat to water as a common resource.

There is no doubt that the Polaris Institute had a major influence on the emergence of bottled water as an issue, and that its institutional context and practices shaped many of the durable framings that were developed to contest markets and the industry in many other places. In 2006 Sustain, a UK food and agricultural agency, released an anti–bottled water booklet titled *Have You Bottled It?* (Wanctin, Dalmeny, and Longfield 2006), which set out several facts about bottled water that resonated with the Inside the Bottle campaign. This was one of many anti–bottled water campaigns that sprang up from 2005 on in the United Kingdom and United States. Meanwhile, Polaris kept up its investigative reporting on all aspects of the industry. In 2009 it published a report on the poor health and lack of environmental regulations in the bottled water industry (Polaris Institute 2009). Later the same year it published *Bottled Watergate* (Cressy, Polaris Institute, and CUPE Nova Scotia 2009). In August 2009 the BBC was accused of wasting £406,000 of public money on bottled water (McCarthy 2009). By 2010, Canada was holding its first Bottled Water–Free Day.

As Marres (2007) argues, issue formation is a performative process, and the associations mobilized in the enactment of a public controversy are partly constitutive of the issue. Detailed empirical research into the beverage industry, involvement in wider global water politics, targeted campaigns focused on already interested or potentially affected constituencies, debates, lobbying, and boycotts were the practical ways in which Polaris defined and enacted a "politics of bottled water." While beverage companies may like to dismiss this activism as the work of unrepresentative troublemakers, the evidence amassed here shows that in postrepresentative politics, when governments are often seen to be weak or failing, issues play a crucial role in articulating what should count as common in democracies. While idealist notions of democracy focus on a general notion of the common good, issue networks demonstrate how that commons is practically enacted or threatened. The effect of Polaris's work was to recruit bottled water to the bigger question of what the place of water should be in the realization of collective benefit and justice. Their answers gave bottles a very bad reputation, and made them into political events and objects.

From Information to Affect: Brita's FilterForGood Campaign

In this section we turn from the work of NGOs and the emergence of anti–bottled water issue networks to a startling advertising campaign for Brita water filters called FilterForGood (Brita 2008). Our analysis of the Polaris Institute's campaign showed how the framing of bottled water as a political issue depended on publicity to "market" the issue and build attachment to it. Here we investigate an advertising campaign whose intention was to promote a commodity rather than an issue, and to build a market rather than a public. The paradox is that this campaign was uncannily similar to anti–bottled water activism, and explicitly referenced it in the marketing of water filters. In suggesting that consumers buy filters, the campaign's rationale was not that filters improved water quality—the explicit function of the commodity—but that this choice would avoid wasteful, single-use, petrochemically dependent plastic bottles. The water filter was made calculable *in relation* to PET bottles, and the basis of this calculation was environmental and ethical.

It is possible to dismiss Brita's FilterForGood campaign as an opportunistic exploitation of existing concerns about bottled water in order to sell water filters; as very canny product positioning and corporate manipulation of consumer passions, driven by the relentless search for competitive advantage. This was certainly a key element in this inventive campaign, but it was not the only one. To reduce advertising to a singular logic that is antithetical to explorations of a common good, or imaginative political invention, denies the complexities of advertising and marketing and the ways in which they can become imbricated with publics. In the Brita campaign, the promotion of water filters generated one of the most potent and shocking attacks on bottled water to date. And the most striking thing about this campaign was the way in which the ads used affect to mobilize consumers and concerned publics in the same moment. The value of this example is that it highlights how issue networks and matters of concern can be extended and elaborated in advertising—and how, even though the action proposed to address this concern is an alternative product, in the gesture of turning to the filter and rejecting the bottle, the consumer is participating in a public, not just a market.

The campaign was launched with a full-page magazine ad in the United States in May 2007. The image in figure 7.3 was disturbingly ambiguous: was it activism or advertising, or both? The representation of someone literally drinking oil was shocking and disgusting, producing immediate visceral effects. The text, on the other hand, was blunt and matter-of-fact. Statistics documenting the amount of oil used to make plastic water bottles contextualized the image as a representation of a technoscientific issue. And what of the website address? Where was the viewer being directed? The automatic assumption was that it was to one of the numerous anti–bottled water sites that had proliferated over the past few years. FilterFor-Good implied an ethical encounter, an invitation to consumers to make a virtuous choice rather than an environmentally exploitative one. Yet the

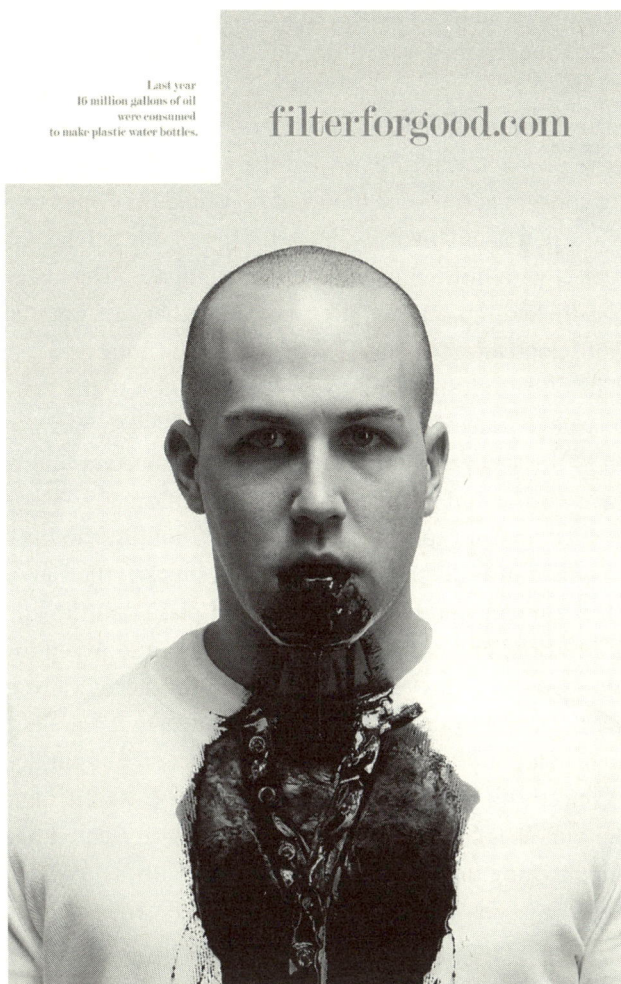

Last year
16 million gallons of oil
were consumed
to make plastic water bottles.

filterforgood.com

Figure 7.3
The Brita FilterForGood campaign, 2008.
Source: Brita GmBH.

site to which viewers were directed turned out to be supported by the Brita water filter company. It offered detailed information on the shocking environmental impacts of bottles, a series of links to mainstream press articles about plastic hazards and wastes, an invitation to "take the pledge" and commit to using filtered water rather than bottles, and consumer advice on why Brita filters were a more sustainable drinking choice—in other words, it was just another infomercial (Brita 2008). This disappointed the expectations of viewers, who thought they were being directed to a powerful activist campaign.[5]

The ambiguity surrounding the origins and intentions of this image was central to its affective power. It explicitly invited the viewer to view bottled water as a matter of concern at the same time as it suggested they buy water filters. By exposing the petroleum intensity of plastic, the ad showed the bottle as a troubling object with disturbing environmental impacts. The imaginary life-worlds constructed in bottled water advertising that cast water as "pure," "organic," or "cloud juice" collapsed in the face of the brute reality of the packaging and its filthy industrial origins. In contravention to the semiotic adjacency of plastic to water in so many bottled water ads—so transparent, clean, and fluid—this campaign brilliantly crafted the consumption of bottled water as the excessive and relentless consumption of *oil*.

In this way, Brita's campaign explicitly called a public, not just a market, into being. Using a powerful aesthetic of disturbance, it shocked the viewer into confronting the fundamental interdependency of plastic bottles and the petrochemical industry, one of the worst polluters around. Accepting this mode of address by giving it even minimal attention meant viewers were drawn into participating in the discursive and political dynamics of a public through their voluntary identification with an issue that appealed to a disinterested concern for a broader environmental good. At the same time, however, the viewer felt the pull of promotion, the mediated discourses of marketing inviting them to choose filters over bottles. Just from looking at this image it was very hard to determine where a market stopped and a public began, for the presentation of the bottle as oil was engaging both these collective identities at once.

How, then, to make sense of this advertising campaign as another manifestation of bottled water as a political issue? In what ways could it be considered a form of market contestation, and not just marketing? And how did it work practically to mobilize public concern about bottled water? The first thing that must be acknowledged about the FilterForGood campaign is that it highlighted the significance of advertising as a powerful affective technology in media-intensive cultures capable of calling various evanescent collectives into being: markets, publics, fans and stakeholders. Advertising is both a market device and an important element in the expansion of publicity across all aspects of everyday life. As Michael Warner (2002) argues, in cultures structured by mass media and consumption, regimes of publicity are omnipresent. This means that the realm of politics is often accessed in the same way as the circulation of commodities: through multiple sites of publicity that make up heterogeneous and dispersed public spheres. Although different sites of publicity involve distinct conventions and techniques (news headlines, advertisements, blogs, activism), each is also capable of illuminating the others through the use of common devices for capturing and sustaining mass attention.

The Brita ad highlighted the complex interarticulations between market devices and public engagement, and the way in which advertising was central to this interworking. It was a potent example of how existing activist publicity surrounding an issue could be exploited to create new and expansive circulation for commodities—and, conversely, how critical reflection on consumption practices within advertising could generate new forms of public engagement. It was also evidence of how marketing could function as a zone of political experimentation and innovation, as a platform for proliferating matters of concern. In this way, the ad generated a distinct anti–bottled water controversy where questions about consumption, politics, and pollution mattered. However, it generated these questions not through information and empowerment, as in the Polaris campaign, but through the affective force of what Jane Bennett (2004) calls "thing-power." In the image of drinking oil, viewers saw a disturbing example of human and nonhuman intermingling that unsettled cultural boundaries and the intelligibility of social order. This was definitely matter out of place that provoked a very uncomfortable and shocking sense of ontological insecurity.

For Bennett, "thing-power" refers to the specific kinds of materiality that are often obliterated by human practices of objectification and classification, such as market framings. In claiming that "there is an existence peculiar to a thing that is irreducible to the thing's imbrication with human subjectivity" (Bennett 2004, 348), Bennett is not arguing for an essentialized materialism; rather, she is insisting that things have the capacity to assert themselves—that their anterior physicality, their free or aleatory movements, can capture humans, even though humans like to think they have the world of things under control. Recognizing the "thingness" of things is not to deny the dense web of relations in which they are always implicated. It is simply to be open to the power of matter and the possibility that it might have the capacity to suggest different ways of being. The

Brita ad deliberately foregrounded the thing-power of the bottle: in this distinct media marketing assemblage, the package appeared more vividly as oil, as matter and movement entrained in a network of industrial chemical processes. And in this *was* its shock effect: not simply in the exposé of the bottle's origins but in the affective connections that were established between the matter of oil and the drinking body.

But how did this activation of the thing-power of the bottle generate an anti–bottled water public? The Brita ad was a particular form of publicity circulating in the realms of consumer culture. However, its distinct aesthetic of disturbance made it ambiguous for viewers: should they apprehend this publicity as consumers or as concerned citizens? The logics of its attraction were uncertain and disconcerting, for it invited both these identifications. In this way, the ad demonstrated how the contexts of commodities and politics can share the same publicity, how they can use the same promotional devices. Even though the logics of consumption presume a clear distinction between individual versus collective identifications, this distinction is difficult to sustain. Consumption is limned as the heartland of individualism: it appeals to the self-motivations of choice and market-mediated identity. These personal choices, however, always imagine a wider constituency, a collective witnessing of our choices and symbolic identifications as a form of public display (Warner 2002, 170). In this way, responding to the political reverberations of the Brita ad, acknowledging its affective address, required imagining a wider constituency of other concerned strangers who were also troubled by the same sense of ontological disturbance.

What this response reveals is the processual and emergent nature of public formation—the ways in which advertising as affect can open up a space for a potential or "infrapublic" (Hawkins 2011, 551) to be politically engaged. The idea of an infrapublic challenges the reification of publics— the assumption that they already exist and are waiting to be convinced by the appeal of reason, or that they are coherent collectives that share a common conviction. Rather, the infrapublic makes visible a different form of political process produced not by means of organized contestation and the logic of critique but through the eruption of different engagements with the world.

The Brita ad was an event in the sense that it used ontological disturbance to create a force field of new relations and problematizations around bottles, and to mobilize diverse publics of affected persons. This acknowledgment of the diversity of publics is crucial. Whereas many viewers of this ad were no doubt open to being affected, thanks to their awareness of activism and its damage to the bottle's reputation, others would have been completely unaware that bottled water was an issue. They responded instead to the reverberation of affect—to the sense of being captured by the image and unsettled by it. This infrapublic was different from an issue public in the sense that it was a fleeting temporal assemblage of affected persons "always in the process of forming and dissolving" (Bennett 2007, 144).

What Brita did was elaborate the problem of bottled water in a new regime of publicity, advertising. In this particular regime, the devices used to extend the issue were primarily aesthetic and affective. Unlike the issue publics mobilized by Polaris, Brita's publics were called into being by the forces of affect rather than by more standard forms of political persuasion, such as critiques of corporations or appeals to universal principles. And although statistical facts were an element in the Brita design, the focus was overwhelmingly on the horror of contamination: a body and a world poisoned by oil. This visual technique highlighted the role of affect and subliminal sensation as important elements in marketing *and* political process. While these elements are acknowledged as central to the operations of advertising, the general assumption is that they are inherently manipulative and dangerous. In contrast, politics and public communication are too often framed as the realm of reason, facts, and careful deliberation. Affect and thing-power disrupt this opposition by showing how material things can bite back and unsettle in ways that produce flux, intensities, and changed awareness. This affective response does not necessarily generate studied political calculations or actions, but it does generate ontological interference—a sense that other realities are possible and present.

The concepts of infrapublic and affect show why Brita's FilterForGood campaign cannot be dismissed merely as canny marketing and manipulation. Of course, the objective of the advertising strategy was to propose a market solution to the problem of bottled water, to suggest a different form of consumption as an ethical way to act. But the fact remains that this market was a hybrid assemblage: those who joined it were affected by the advertising as concerned subjects, and their consumer choices were both personal and public. What the FilterForGood campaign shows is how issues circulate in the public domain—their potential to move and launch new matters of concern and debate in numerous settings, and to prompt new sites of possibility. In this way, Brita's marketing campaign can be considered a form of political experimentation: the ads modified and played with the issue of bottled water so that it became connected to the wider politics of oil and disposability. The devices of marketing represented the plastic bottle as a thing with the capacity to shock and disturb. This representation was aesthetic and political; it conferred agency on the packaging by foregrounding its thing-power. Although this agency may not have prompted the type of activist subject targeted by Polaris—someone ready to sign petitions and lobby—it did prompt a form of market refusal or boycotting of bottled water. It also prompted recognition that affective responses can change awareness and generate inchoate and more-than-human publics.

However, while the Brita ad experimented with the affect and materiality of packaging to undermine bottled water markets, it also implicitly supported those very same markets by endorsing their challenge to the tap. In advocating the use of water filters, FilterForGood amplified public uncertainty about the quality of tap water that bottled water markets had actively fueled. Though the filter was framed as more sustainable than the bottle, the assumption across both markets was that tap water was suspect. This shows how filters were positioned in relation to the economy of qualities that bottled water markets created *and* to the issue networks and contestation they prompted. Brita is a powerful example of how the complex logics of product positioning and singularization in markets can work in multiple directions, from competitive product relations to emergent publics and activism.

Recognizing these dynamics does not undermine the argument that advertising and affect can prompt the formation of infrapublics and important sites of the political. But it does underscore how markets and advertising seek to contain their political contestation. In presenting its product as environmentally superior, the Brita corporation acted as a spokesperson for the planet. Like NGOs, the corporation was concerned for the environment; however, the ultimate objective of its concern was to build new markets—to make an issue public into a market. In postrepresentative politics, markets are as emergent to issues as issues are to markets. Ultimately, however, Brita's advertising of water filters as a better option than bottles came unstuck as the externalities of this market were exposed by controversy.

Soon after it was launched, the FilterForGood campaign was targeted by *copyranter*, a blog about the advertising industry (*copyranter* 2008). *Copyranter* ran an online poll surveying attitudes toward the drinking oil campaign, with 61 percent of respondents voting that it was very effective and 38 percent deeming it hypocritical, claiming corporations can't be trusted as authentic political spokespersons. This poll sparked an organized backlash against Brita, which was attacked for producing bulky plastic filters that were nonrecyclable and that, just like bottles, ended up in landfill. A Take Back the Filter campaign was launched in Canada, based on Food and Water Watch's Take Back the Tap campaign advocating for clean and affordable public water (Take Back the Filter 2008). This issue network lobbied for Brita and other water filter companies to set up effective recycling programs for water filter cartridges. In November 2008, Brita announced it had teamed up with Preserve, a company that made recycled plastic products. This partnership would not only recycle its filters but would also restore and strengthen Brita's green credentials. The recycling program began in 2009.

In the face of this controversy, Brita didn't back down from its FilterFor-Good strategy. Instead, it elaborated and extended it. Although the drinking oil images were pulled, several other videos were launched on a special YouTube channel that developed the theme of plastic waste as a major global environmental problem. An Eco-Challenge prize was also set up inviting university students to devise a new anti–bottled water campaign; like Polaris's following, this constituency was deemed the most responsive to help elaborate the issue. The multidimensional use of print, television, and online media, viral videos, and celebrity endorsements extended the reach of the campaign. Again, in a borrowing from Polaris and other issue networks, Brita invited people to post links, sign up to "take the anti–bottled water pledge," and organize events at schools and universities—using Brita water filters, of course. In May 2010 Brita launched yet another video ad, *The Earth Needs Brita* (Brita 2010). Although the drinking oil ads had been pilloried as politically hypocritical, this corporate concern for the Earth went unnoticed. The issue, the market, the corporation, and the public concern had become thoroughly mixed up and almost indistinguishable. Bottled water was bad, and Brita had become one of the most effective promoters of this contestation.

Do Something: Bringing Back the Water Fountain

In our final example, we look at a very different approach to contesting bottled water markets. Although this example draws on many of the framings established by the Polaris Institute's Inside the Bottle campaign and Brita's references to disposable plastic as resource-intensive waste and pollution, it approached the contestation of markets with a strategy focused on extending access to free water in public spaces. As we showed in chapter 3, many bottled water markets are built through promoting constant sipping as an essential part of self-care and health regulation. Carrying or buying bottles of water while on the move is a necessary part of this practice, so that bottled water comes to function almost as a personal mobile emergency supply. This market and this consumption practice, evident in everything from vending machines to special pockets in backpacks for carrying the bottle, have now been recognized as major contributors to growing urban waste problems. Garbage bins overflowing with empty bottles have been blamed for destroying urban amenity and challenging the capacity of waste-management infrastructure to keep up with the discarding of disposable items in public space.

In this context, the Bottled Water Alliance emerged in Sydney, Australia, in 2008. The alliance was organized by Do Something, a nonprofit organization established by activists from the environmental group Planet Ark and a former executive director of the National Trust of Australia. The overarching aim of Do Something was to combine "the resources of the business community, government, and the goodwill of the Australian public to promote positive social and environmental change." Other campaigns focused on reducing food waste, reducing paper wastage in workplaces, and banning nonbiodegradable plastic bags. Bottled water was an obvious addition to this repertoire of issues (see figure 7.4) (Gruden 2008).

In establishing the Bottled Water Alliance, Do Something invoked many of the framings of the issue in play since Polaris's 2005 Inside the Bottle campaign: price gouging, the waste and environmental burden of plastic bottles, and exploitation of aquifers. To these it added climate change, which had become a major global concern by 2009. Making bottles out of a nonrenewable resource, oil, and transporting them around the world using more nonrenewable resources was contributing to greenhouse gas emissions and was difficult to justify in the face of Sydney's excellent tap water quality. In devising the campaign, a group of partners was assembled to promote and support it. The group included "founding sponsor and partner Culligan Water," an international water filtration company, local councils, green magazines, and a scattering of B-list celebrities. Unlike Polaris, Do Something had no qualms about corporate or government sponsorship

Figure 7.4
The Do Something website.
Source: Do Something.

and partnerships, which were essential to its financial base, and to the management and development of the issue.

The Bottled Water Alliance had a number of activities and objectives,
all concentrated on the explicit goal to "reduce bottled water sales in Australia by 20% by 2010" (Bottled Water Alliance 2008). Rather than document all the ways in which the issue was developed, here we examine one
strategy that stands out as perhaps the most effective: the Manly Council
Water Fountain Project, in the seaside suburb of Manly, north of Sydney's
central business district, which aimed to reinvent the water fountain in
public space and provide a free alternative to the purchase and carrying of
single-serve bottles of water (Manly Council 2009, 2010). In October 2007,
the Manly Council delivered a presentation to the Local Government Association (LGA) on the increasing problem of bottled water consumption and
its waste impacts. Growing numbers of pedestrians were purchasing bottled
water on the move and rapidly discarding the empty containers. The emergence of this new consumption and disposal practice was evidence of a

growing market segment concerned with making healthy choices, and with carrying and constantly sipping water. Council research showed that rapid growth in PET waste in public spaces was a direct effect of the increased consumption of single-use bottles of water. While fast-food packaging had always been a major litter problem, escalating bottled water use was making it far worse and far harder to manage.

The council's research also showed that many pedestrians disliked using public water fountains because they were often dirty or vandalized, or because the water tasted unpleasant. Even though the water was of high quality on hot days, it tasted warm and very chlorinated. In response to this research, the Manly Council partnered with Do Something and the Bottled Water Alliance to install a new style of water fountain that was appealing to pedestrians and offered a real alternative to single-use purchasing of water. In 2008 the first water fountains were installed along Manly's heavily used pedestrian mall, The Corso. These water fountains, designed and provided by Culligan Water, were state-of-the-art designs that were vandal-proof, wheelchair-friendly, and delivered cool filtered water. They were also designed to facilitate easy refilling of pedestrians' own bottles. In launching the new water fountains, the Manly Council installed extensive signage inviting pedestrians to taste the cool, filtered water, reduce plastic waste, and combat climate change. Drinking from these new public things was framed as a virtuous act, a personal gesture that made a difference to the environment (see figure 7.5).

There is no question that the trend of mobile eating and drinking has made the maintenance of urban amenity much more difficult. In this context, the plastic bottle of water has become a major target. It has been criticized as not only the cause of increased litter but also as a powerful indicator of the loss of public access to water. In many other cities around the world apart from Sydney, anti–bottled water campaigns have developed that show up the madness of individuals' buying and carrying water when water fountains once met their hydration needs much more efficiently, and with no waste. Many of these campaigns attack the carrying of bottles as evidence of the decline of "public things" (Honig and Gelonesi 2013). In contrast to the bottle of water as a singularized commodity for individual use, the water fountain is a shared resource. Rather than carrying a private water supply, drinkers approach the water fountain as something they use in common. In stooping to drink, they show how public space is made

Figure 7.5
The Do Something water fountain campaign at The Corso, Manly, Australia.
Source: Do Something.

public by particular objects and the connections that are established among all who use them. Bonnie Honig's argument is that when two people have an affective relationship to the same public thing, they also implicitly have a relationship to each other. In this way, a new carrying practice has been powerfully criticized as a threat to the value and role of public things as critical resources for democracy.

Some anti–bottled water campaigns defensively attack the privatization of water as a sign of relentless neoliberalism and the corporate takeover of everything. However, this framing of the issue seems less effective than those campaigns that positively reassert the value of the water fountain as a way to reimagine public space, ethical drinking, and being in common. This was the implicit focus of the Manly Council campaign. What was so interesting about this campaign was the way it sought to challenge a consumer carrying practice and displace the bottle by *revaluing* public water and making it available in a new form. This intervention worked at the levels of infrastructure, materiality, and affordances—as well as at the levels of issues and democratic ideals. It sought to intervene in market definitions of convenience and mobile consumption by creating new arrangements that enabled alternative forms of access to water on the move. In other words, the water fountain sought to compete with the bottle on its own terms.

The Manly Council water fountain project was enormously successful, and has been copied in many Australian cities. By reinventing the water fountain, a public thing contested the rise of a new market in water and a growing urban waste burden. This example also shows how markets and new consumer practices are contingent on complex sociotechnical networks and relations that can be challenged. The Manly Council recognized consumers' new attachments to water and their concern for health and hydration, but it offered the public supply as a way to address this need, as another choice. By providing an alternative to purchasing and carrying water, the water fountain helped make public space a space in common, not just a space of commercial interests and private consumption. It also reasserted the fact that public space is made real through material things that are durable and shared—things that are accessed by many other strangers. Through good fountain design, the fear of drinking from a shared supply— of making contagious contact with strangers—was eliminated. Instead, the water flowed freely into the mouths of strangers, and this common flow implicitly linked all who drank it into a public.

Carrying a plastic bottle doesn't just make water into a private possession, it also makes it into a personal responsibility. Many of the new water fountains included a refill station that recognized people's desire to carry water about in bottles and be constantly hydrated. In this way, a "need" that has been created by the affordances of bottled water markets and health advertising was explicitly accommodated and catered to by changed public design. The public form of water provision was shaped by the rise of a market and new drinking practices at the same time that this public provision sought to undermine this market.

The water fountain project showed how the provision of a new public device changed relations between and among pedestrians, water, and drinking habits. Just as bottles helped to format new drinking practices, so too did water fountains. In this case, they exposed the bottle as an unnecessary, expensive, and weighty accessory and instead offered access to water that was outside of commercial culture—water that was free, public, and shared. By stooping to take a drink at the water fountain, pedestrians enacted a public rather than a consumer disposition, and were reminded of the role of water in sociality and being in common.

Conclusion

This chapter has focused on the practical techniques and devices used to contest bottled water markets and turn the commodity into an issue. Rather than assume that there is a singular "politics of bottled water" based on a reductionist critique of the evils of markets and corporations, we have

Figure 7.6
The Do Something website, 2013.
Source: Do Something.

offered up the three examples to show how markets become caught up in reflexive processes of issue formation and controversy that expose their contingencies. Central to these processes is a focus on the externalities of markets, an exploration of all the relationships and interconnections in which the market is embedded beyond its explicit frames. And just as these relationships are multiple, so too are the possibilities for the political. As the Inside the Bottle, FilterForGood and Do Something campaigns all show, bottles prompt an immense variety of concerns. They also prompt an immense variety of activist practices, devices, and techniques that underscore how issues are reformatting politics as postrepresentative, experimental, affective, networked, and material.

Despite their differences, the overarching commonality of these campaigns is the potency of water as a biopolitical material. Positioning the rise of the bottle in wider networks of water provision *as a disruptive event* has been exceptionally effective because everyone is entangled with water, everyone drinks, everyone is potentially affected. Of course, this doesn't mean that everyone becomes involved; rather, part of the power of the issue emerges from an implicit appeal to the ordinary act of drinking water as a universal and common experience. The rise of the bottle invited reflection on that taken-for-granted experience because this new mode of water delivery made visible a vastly different approach to organizing supply: from safe public flows to containerized, branded, expensive single serves.

The three examples investigated here also complicate easy distinctions between the public and the market. The emergence of markets and publics involves practices of formatting and framing, both of which depend on devices and techniques to capture elusive subjects. And they can often operate symbiotically or dialogically, harnessing and poaching from each other to achieve their own ends. This becomes really evident when one considers sociomateriality, everyday habits, and the expectations and norms of living to which they give rise. Public provision is just as susceptible to changing regimes of living generated by market commodities—such as drinking branded water because it is "purer"—as markets are susceptible to emergent forms of public concern that have been mobilized by issue networks.

As we have argued here, the rise of issue networks and controversies around bottled water was an event that triggered a proliferation of new relations, problematizations, and identities for this once innocent commodity. For beverage companies, this reputational damage led to a wide range of strategies to contain the issue and reestablish positive qualities for the bottle and the water and the market. We investigate these responses in detail in the next chapter. For consumers who became attached to the issue, who felt implicated in the controversy, reputational damage was experienced as discomfort with the bottle. It was now something with the capacity to prompt forms of political calculation and action, something that posed questions to them and invited them to enact a public rather than a consumer drinking disposition. Whether this was done through boycotting or signing petitions or buying a reusable bottle or feeling horrified at the idea of drinking oil or stooping at the water fountain is not the only question. The main thing is that in these responses, the event of bottled water as an issue reverberated across multiple sites and inaugurated various forms of ethical drinking.

8 Spinning the Bottle: Ethically Branded Bottled Water and the Rise of Cause-Related Marketing

Just as anti–bottled water activism and issue networks can be considered an event, so too can the rapid growth of ethically branded bottled waters. Around the same time that bottled water markets were coming under attack on numerous fronts, consumers in the United Kingdom were confronted with coolers full of Thirst Aid Water and One Water, while in Australia, striking pink-capped bottles of Mount Franklin, Australia's market leader, hit the shelves during Breast Cancer Awareness Month. In the face of intense negative publicity, some bottled waters were being requalified as agents of positive social change. Buying Thirst Aid Water meant a percentage of profits would be directed to clean water projects in Africa or Asia. Participating in Mount Franklin's Pink Lids campaign was meant not only to prompt women to have a mammogram but also to remind consumers of the generosity of the corporation, Coca-Cola Amatil, in supporting breast cancer care–related initiatives. Here were bottled water markets supporting issues rather than being issues; here were bottled water markets making various matters of concern visible targets of economic action rather than excluding them as troubling or irrelevant externalities.

In seeking to understand the rise of niche markets in ethically branded bottled water and cause-related marketing (CRM) from the big beverage companies—two similar but distinct strategies—it is easy to assume a neat logic of causality. Markets come under attack, the product suffers reputational damage, so the PR officers and issue managers are wheeled in to reposition and redeem the product in the interests of market protection and competitive strategy. Critics have been quick to label the rise of ethical concerns in markets as nothing more than a corporate counterdiscourse, or "greenwashing." From this perspective, CRM and ethical branding are deliberate strategies designed to obscure the harsh realities of capitalism and its fundamental logic of profit and exploitation.

This chapter resists such an analysis. Although it is impossible to deny a relationship between issue networks and the rise of ethically branded water, our aim is to understand the dynamics of this relationship beyond teleological or conspiratorial assumptions. The event of ethical water and CRM is not simply the result of identifiable causes unfolding in a linear pattern of action and reaction, or opposition and assimilation. Rather, it reflects the complex dynamics of emergent causation within market assemblages. As discussed in chapter 1, emergent causation refers to the processes whereby forces and agency emerge and circulate, and become implicated in further changes through feedback loops. These forces are not simply external, in the form of explicit market contestation, for example; they are also part of the way in which market assemblages are always reflexively coordinating the multidimensional relations in which they are caught up, and their diverse practices of calculation.

As bottled water becomes a political object, it acquires the capacity for new forms of expressivity that extend across multiple sites—not just those designated as activism. Its material semiotics becomes denser and more layered through the accumulation of new affects and information permeating market assemblages. The challenge is to investigate how these political affects, information, and other realities become implicated in the reflexive organization of the market. If activism and issue networks have made bottled water into a matter of concern in some places, then the question is, how does this political event enter into the calculative processes of markets, and how does it come to mediate relations between consumers and producers? Ethically branded bottled water sold in niche markets and the promotion of causes by global beverage companies are two key developments that offer some answers to these questions. They signal the emergence of what Andrew Barry (2004) describes as "ethical capitalism," and, following Barry, we want to take the corporate concern with ethics seriously by investigating the origins and implications of "ethical capitalism" in some bottled water markets as a distinct development. For Barry, "ethical capitalism refers not to a system, but to a set of ways of acting on the conduct of business activity" (196). This resonates with Michel Callon's (2009) account of how some markets are "civilized." Using the example of carbon trading, Callon sees attempts to render markets responsive to

various issues and problematizations as a specific sociotechnical development. Rather than investigate these developments as yet more evidence
of neoliberal claims that markets can solve everything, Callon, like Barry,
insists on understanding the particular processes whereby markets render
controversial events debatable *and* as resources for experimentation in new
market forms. These processes reveal specific forms of market reflexivity:
the capacity to design devices and experimental networks that renegotiate
the boundaries between economics, culture, and politics.

Just as we explored in chapter 7 the practical devices and techniques
whereby issues and publics are formulated to contest bottled water markets,
so we now approach ethical branding and CRM. The aim is to understand
what methods are used to make a bottled water brand and market "ethical."
How does a corporation make visible or demonstrate its ethical concerns?
And in what ways do these practices shape new relations between products
and consumers that generate various forms of ethical value or capacity in
the bottle of water, the drinker, the brand, and the beverage company?
Attention to the processes by which a market develops specific calculative
capacities to enact ethical forms of action and value is only part of our
focus, however. We are also concerned to understand how the event of bottled water activism—the emergence of significant critique and opposition,
as examined in the previous chapter—is implicated in these developments.
In what sense do forms of market contestation and controversy cease to be
externalities and instead become unavoidable actants or elements in the
architecture and calculative equipment of a market?

One way to begin to unravel the interactions between anti–bottled water
issue networks and ethical branding is to consider their specific orienta-
tions to biopolitical framings of bottled water. As we saw in chapter 7, many
activist campaigns frame bottle water as a significant threat to the provision of water as a safe public resource. Making water into a market thing
undermines trust in public utilities and the quality of tap water, fuels environmental pollution, and "rips off" consumers with prices thousands of
times higher than the cost of tap water. In ethical branding, the bottle is
also often positioned in relation to wider biopolitical regimes, but here the
relationship is presented as positive rather than destructive or deceptive.
"Ethical" bottled water does not generate disturbing ontological interference but rather ethical connections and new political capacities. Buying
these brands can help ameliorate anything from water scarcity in Africa to

breast cancer to states of emergency such as natural disasters. So biopolitical framings persist, and the challenge is to understand how they are specifically enacted when a business or corporation, rather than an NGO or issue network, becomes their spokesperson, and when that corporation seeks to highlight rather than suppress bottled water's biopolitical associations. If biopolitics involves broader concerns with water and the sustenance of life, how do markets and brands deal with the challenge of demonstrating their ethical actions both to the consumer and to the often remote sites where serious water issues play out at a distance from the consuming majority (Chatterjee 2004)?

In making these points of comparison, we aim to show how qualifying the bottle of water as ethically implicated in a multitude of relations and networks beyond the immediate market frame is a practical challenge confronted by both business and NGOs. CRM and the ethical branding of bottled water have to be seen as both new forms of market assemblage and as related to activism and market contestation, not in opposition to them.

To understand how some bottled water markets have experimented with practices of "ethicalization," we explore three examples in this chapter. In the first, we examine the rise of ethically branded bottled water. How do these niche brands, and the markets they help organize, reframe issues and causes as critical actants in market arrangements, and what are the effects of this distinct market organization? In the second example, we investigate CRM by the big beverage companies, with a particular focus on the Mount Franklin brand in Australia. What causes are identified as appropriate to demonstrate the ethical qualities and concerns of Mount Franklin? How does the brand as interface enable such exchanges between markets and causes? And how do these causes become implicated in the reconfiguration of market assemblages? In the final example, we look at how Coca-Cola Amatil (the manufacturers of Mount Franklin) responded to an aggressive market attack in some remote Australian indigenous communities. When the sweetened beverages they were selling in community stores were implicated in higher rates of diabetes and other serious public health issues, an alternative product in their beverage range, bottled water, was advanced as a positive solution. How was bottled water positioned as an emergency technology during this crisis, and how did a corporation become identified as an effective participant in addressing serious indigenous people's health problems?

Ethically Branded Bottled Water: Making Not-for-Profit Markets

In April 2008, the industry journal *Beverage Daily* reported a bright light amid a downturn in the UK bottled water market: the rise of "ethical brands" of bottled water. Even as the major bottled water brands were feeling the impact of a growing consumer backlash, this small segment of the bottled water market was growing apace, doubling its sales volume in a few short years (Merrett 2008). Following the lead of Belu water, launched in 2004, a range of ethical brands, including One Water, Frank Water, Thirst Aid Water, and Thirsty Planet, appeared in the United Kingdom. Although overall sales of bottled water in the United Kingdom dropped by 11 percent between 2006 and 2008, sales of ethically branded water in the same market grew by 23 percent the following year (Bainbridge 2009; Merrett 2008).

What was distinctive about most of these new ethical brands was their focus on donating a percentage of sales to clean water projects in Africa and Asia. As *Beverage Daily* wrote, these products were "beyond the focus of developing markets," committed instead to "providing charitable assistance" to others in need (Merrett 2008). In the same period, the United States and Australia also saw not-for-profit ethical waters enter the bottled water market. Ethos Water, established in 2001 and sold through Starbucks in the United States, provided the model for Give Water (also U.S.-based), which, like Ethos, provided a portion of sales to charitable causes. Thankyou Water entered the Australian bottled water market in 2008 and was followed by the Australian launch of the UK brand One Water in 2009. Both these products channeled the entirety of their profits to clean water projects in Africa and Asia. One Water Australia, however, has recently initiated a local program of "water security" projects in remote indigenous Australian communities (One Water Australia 2013).

As we have indicated, it is important to distinguish between ethically branded bottled water and CRM, initiatives considered in the next section. The self-proclaimed not-for-profit ethically branded waters are niche products generally marketed through a "grassroots" narrative of social justice, reminiscent of many NGO campaigns. Forsaking the usual rationale of markets, the bottle of water is explicitly publicized as something that has been created to do good rather than make money. The ethical imperative is not an occasional add-on or passing cause but the central rationale for the development of the product. These products usually emerge from smaller companies started by people who have a clear moral agenda and who want to connect with ethical consumers: a niche product for a niche group of drinkers. This is exemplified on the Thirsty Planet website in a narrative echoed across the ethically branded bottled water market:

> We are lucky. In the UK we have access to clean safe water at home, at work, at school. When we're out and about we can buy bottled water easily. Yet in developing countries over a billion people still cannot access clean water. By supporting Thirsty Planet and choosing to drink our water not only will you enjoy a quality British spring water but you will help save and change the lives of millions of African people who are not as fortunate. Buy a bottle, change a life! Doing something truly amazing really doesn't get any easier. (Thirsty Planet 2007)

Ian Robinson (2007) describes these businesses as "moralized firms" because of the central role ethical imperatives and calculations play in both the producer and the consumer.

Many of these businesses and brands depend on the investment and public patronage of diverse business and cultural elites rather than well-known beverage corporations. Although their market share is very small compared with that of the big bottled water brands, they represent an interesting market arrangement in which the production and purchase of ethically branded products are figured as a form of philanthropy in the name of global water justice. Although the big beverage companies often invite consumers to reflect on their personal hydration levels as a form of ethical self-regulation, these brands invite the consumer to reflect on the global realities of thirst and water scarcity. In this way, biopolitical questions about the place of water in "how to live" (Collier and Lakoff 2005, 33) invite very different answers. In reaching for the ethically branded bottle, the consumer is acting at a distance in relation to global water issues. In contrast, reaching for one of the big beverage companies' bottles is usually an intimate ethical action directed at care of the self and not the other (see figure 8.1).

But how does the consumer understand the purchasing of these niche bottled waters as an ethical action? For Aihwa Ong (2006, 501), the capacity of the consumer to recognize and respond to ethical branding is characteristic of new citizenship forms, in which "the security of citizens, their well-being and quality of life, are increasingly dependent on their own capacities as free individuals to confront globalized insecurities by making calculations and investments in their lives." This form of citizenship—or "consumer citizenship," as it is often referred to—locates ethical action in

Figure 8.1
One Water advertisement in a shopping center.
Source: Global Ethics (Australia).

market choices. However, markets shape these choices in different ways. What ethically branded water does is link wider biopolitical concerns about global water security to the specificity of markets that are explicitly designed to offer a partial solution. These markets acknowledge many of the concerns that activism has raised about bottled water's implication in water security and privatization. However, rather than boycott or refuse the bottle as a political action, they reconfigure and redeem it so that choosing it becomes politically legitimate. In this way, ethically branded water seeks to explicitly render consumer choice a moral rather than a purely individual calculation. Consumption is reconfigured as an expression of virtue and concern for others (see figure 8.2) (Foster 2008, 227). Along these lines, the founder and CEO of One Water, Duncan Goose, has stated, "People have recognized that water is water; why wouldn't you opt to buy a brand that changes people's lives?" (Fry 2007).

Implicit in this "consumer-led" activism is the absence of any *extra* work required by the bottled water drinker. In the act of choosing this brand

Figure 8.2
"When you drink One, the world drinks too," 2013.
Source: Global Ethics (Australia).

rather than another, ethical work is undertaken through ordinary consumption. But it is not just a matter of consumers making the right choice; it is also a matter of establishing that the bottle of water is a product of a market focused on social good rather than on capital growth and purely commercial calculations. Duncan Goose has explained this in relation to One Water's launch in Australia: "We are not asking people to give money or trying to grow the Australian bottled water market, we are just asking current buyers to make an educated purchase and simply choose *One* in the knowledge that as they drink water, so does Africa and Asia' (Palmer 2009). Through marketing that conflates waters from radically different sources (bottled water in Australia, pumped water in Asia) and thus equalizes all water ("water is water"), choosing the ethical brand of bottled water is presented as an act of generosity and connection rather than as an expression of preference for a niche consumer product. These ethical brands enable consumers to make a political gesture without effort *and* without explicitly identifying with an activist counterpublic; these gestures also offer translocal connections and scale shifting: choosing here reverberates there.

Ethically branded bottled water products also rely on networks of expertise, cultural status, and association to enhance the impact of the individual story and to build brand value. Marketing often involves disseminating statistical data on water scarcity and images of impoverished African villages, celebrity endorsements, and the participation of other crucial actors such as NGOs and business sponsors. These devices and agents are called on to help build an ethical economy of qualities for the bottled water. One Water in the United Kingdom, for example, prominently displays its celebrity ambassadors on its website and in its advertising campaigns. The Hollywood star Mischa Barton and the UK actors David Tennant and Claire Goose, sister of One Water founder Duncan, call on bottled water consumers to "Switch for Africa."

In the case of ethical brands, while these various techniques and devices are designed to build an "ethical" interaction and exchange between the product and the consumer, they also have other implications. Like issue-based activism, they use information and knowledge about water politics to educate or "empower" the consumer, and celebrity endorsements to increase their appeal. However, unlike issue-based campaigns, they invite the consumer to respond to this knowledge by choosing *better* rather than not choosing at all. In this way, consumption of a demonized product becomes acceptable. But in choosing *better*, the consumer remains attached to a market rather than a public, and the ethically branded bottle doesn't challenge the more fundamental practice of drinking water from single-serve disposable containers. An unsustainable market-based drinking practice remains intact, and the paradox of using the proceeds of this practice to develop basic water infrastructure elsewhere is striking. Ethical brands of bottled water gather around the same political concerns as anti–bottled water activism, even to the point of declaring bottled water a problematic venture. In an interview about Thankyou Water's origins, Daniel Flynn declared bottled water "a little silly": "Why pay two to three dollars for something that we can all get from the tap for free?" (Prescott 2012). The opportunity to leverage bottled water's success to do good work apparently neutralizes these concerns: "At Thankyou Water we … empower you to play a part in solving the World Water Crisis" (Thankyou Water 2008).

Ethical brands of bottled water—again like anti–bottled water activists—position themselves against the mainstream bottled water brands. Like Brita (see chapter 7), they opportunistically respond to the countermovements brought to the bottled water market by activists and other political actors keen to point out the unsustainable, exploitative nature of the bottled water industry. The inference of the ethical brands we have discussed is that their work is to politicize and reclaim a portion of a problematic market that otherwise operates with profit as its sole logic, to do "charitable" work rather than "develop" markets. Yet—and also as with Brita—other motivations are also at work, often lacking transparency. This was evidenced recently when Thankyou Water came under fire for failing to disclose to consumers the channeling of 30 percent of its profits into missionary work in developing countries (Battersby and Holland 2013). The biopolitics of water takes on another dimension here. The imperative to universalize clean water access is thus highly targeted across a range of motives and by a diverse range of actors in the making of the bottled water event.

Cause-Related Marketing: The Case of Mount Franklin in Australia

In this section, the focus shifts to how large beverage companies have engaged in CRM. CRM is another strategy of ethicalization that involves similar (though also quite distinct) practices to ethical branding. In this case, the brand is not a new invention: it is an already established platform

for the organization of economic action, usually connected to a global corporation. By attaching the brand to various causes, the corporation aims
to make itself visible as ethically and not just economically active, and
to associate the product with "good work," biopolitically and socially. In
this section we examine some of the ways in which beverage corporations
attach themselves to causes, and ask whether these attachments reconfigure market assemblages. The question is, how can we understand and
critically assess the practices of a corporation as demonstrations of ethical
conduct? Or, as Barry (2004, 200) puts it, "How can businesses possibly
provide an environment for ethical forms of action?" We explore these
questions through the example of Australian bottled water brand Mount
Franklin's CRM strategies over the past decade.

CRM campaigns around bottled water gained traction during the early
twenty-first century as part of a broader uptake of corporate social responsibility (CSR) strategies among corporations.[1] While there is no doubt that
this connects to the emergence of bottled water as a problem and an issue,
the prevalence of these campaigns is part of a much more extensive set of
transformations that were already in place in the business sector. The waves
of privatization and public-private "partnerships" during the 1980s and
1990s brought private capital into very public arrangements of civic practice that positioned corporations as social rather than just economic stakeholders. In this context, corporations and other nongovernmental actors
were drawn into regimes of governance and encountered rising expectations to meet both social and environmental concerns.

These general historical changes offer an important background to the
brand and its positioning under analysis in this section: CRM within a CSR
program for the Mount Franklin bottled water brand in Australia. However,
they don't explain how the rise of extensive issue networks and market
contestation becomes implicated in these developments. While there is no
question that many corporations now seek to demonstrate ethical conducts,
beverage companies have faced an exceptional amount of critique and contestation, which might suggest that ethical practices emerge only in response
to significant market pressure. This is a legitimate claim to make in relation
to bottled water; however, as we have argued, it can become dangerously
teleological, too focused on explicit reactive practices rather than the ongoing reflexive performance of the brand. As chapter 7 showed, much anti–
bottled water activism has focused on the externalities of markets, on the

effects and activities of beverage corporations beyond the market frame. This activism has undermined markets and exposed beverage corporations to significant public criticism and "reputational damage," to use the industry language. How, then, does the corporation manage such negative exposure and seek to re-present itself as an ethical agent? How does it seek to reframe the bottle of water as an ethical rather than a political object? How does it seek to develop brand strategies that have what Barry describes as "anti-political" (Barry 2004) effects? In posing these questions, our aim is to investigate how political events such as activism are part of the ongoing process of market assemblage and brand management, part of the way in which markets are always reflexively caught up in both internal and external dynamics.

If the turn to ethics and causes is connected to the political event of bottled water, as well as to other significant historical shifts, the question is what this turn practically involves in relation to the brand and corporation. Barry (2004) argues that internal strategies within a company around practices of CSR are completely inadequate if they do not also generate external effects—if they aren't also promoted as part of the public performance of the company. CRM offers the perfect strategy for publicly performing "social responsibility"—for reframing the corporation and the market as an ethical assemblage. Marketing is, after all, one of the most visible and public devices available to the corporation. As Celia Lury (2004, 22) argues, it is a performative discipline that has become central to the ongoing qualification and requalification of the product. Marketing doesn't simply involve the generation of extensive knowledge and information about the many dimensions of the market, from consumer behavior to competitors; it also underpins the flexibility of the brand as a dynamic platform for the management of change. Marketing information has become the central rationale for processes of product differentiation and for positioning the brand in relation to changing conceptions about the consumer (Lury 2004, 28). Attaching the brand to various causes reflects this, as a brief history of CRM for the Mount Franklin brand in Australia indicates.

Coca-Cola Amatil and Mount Franklin: "Good Works" at CCA

Coca-Cola Amatil (CCA) is Australia's largest producer of nonalcoholic beverages, with a net profit of A$558.4 million in 2012 and a potential consumer base of 270 million across its markets in Australia, Indonesia, New

Zealand, Fiji, Samoa, and Papua New Guinea (CCA 2010). As the Australian licensee of TCCC products, soft drinks are a core part of CCA's beverage range. However, with a continuing decline in the popularity of high-sugar products internationally, noncarbonated beverages are increasingly important to CCA, representing 26 percent of the company's sales in 2012, compared with 5 percent in 2001 (*Australian Financial Review* 2008). Bottled water currently accounts for 15 percent of this share. As elsewhere around the globe, the bottled water market in Australia was tiny twenty-five years ago, comprising mainly boutique, imported brands. The introduction of the Mount Franklin brand by CCA in 1989 marked the beginnings of a local bottled water market, and CCA has subsequently added a variety of still and sparkling bottled waters to its stable. As a result, CCA's stake in the Australian bottled water market was 34 percent in 2012 (CCA 2012).

Of the thousand or so bottled waters on the Australian market, Mount Franklin is the highest-selling and most recognizable brand in the country. Its iconic 600-milliliter bottle is placed prominently in retail coolers around Australia. Unlike its licensed brands, CCA owns Mount Franklin, and consequently controls the product's marketing and distribution nationally. Mount Franklin water has also become the company's most widely utilized product for a host of CSR initiatives, including CRM, focused particularly on "health and well-being." Mount Franklin's stated target consumer base is eighteen- to thirty-five-year-old women, and the bottle's design references this base constituency with shapely contours and a tapered waist. As the product's visibility has increased, however—partly through its CSR program—CCA has extended its marketing of Mount Franklin to a broader constituency, addressed through a targeted appeal to healthy lifestyles and civic-mindedness. Its developing CSR initiatives clearly invite an association between a product that *does* good with one that *is* good for consumers.

According to a CCA spokesperson, the Mount Franklin brand has been "working hard" in "different roles" within the company's CSR program (Potter 2009d). Because of the significant soft drink stable licensed to CCA, growing public concern over the negative health effects of these products shaped CCA's CSR agenda from the early 2000s on. Mount Franklin was increasingly positioned as a health-related product with multiple physical, emotional, and philanthropic benefits. In 2004 a new information website was created for Mount Franklin that offered consumers information on the qualities of natural spring water. In 2005 the tagline "Makes You Feel Good" was added to the product's branding. This was followed in 2006 by the expansion of CCA's explicit association with women's health through the beginning of its Pink Lids initiative.

The Pink Ribbon campaign to raise awareness of and funds for breast cancer research was formally established in the United States in 1992, and is now one of the most prominent targets of corporate sponsorship around the world. In "partnership" with the Australian National Breast Cancer Foundation (NBCF), CCA turned the signature blue lids of its Mount Franklin bottles pink for the month of October (Breast Cancer Awareness Month), donating a portion of sales to the NBCF. Mount Franklin's association with breast cancer research and "positivity" were cemented in 2007 with the launch of the Drink Positive, Think Positive campaign—a new tagline for its branding—and a "Well of Positivity" website. Consumers were invited to leave messages of support to breast cancer sufferers and survivors on this website (see figure 8.3); for every message left (and thus every consumer enrolled), a $1 donation to the NBCF was made.

Breast cancer is still Mount Franklin's most prominent target of CRM, with its Pink Lids campaign repeated in October each year. This is a crucial material strategy, visually linking the product with its good work: here the commodity becomes a benevolent and important tool of public education. A CCA spokesperson put it like this: "You walk into any supermarket or corner store and you see this wall of pink and immediately women think, 'It's breast cancer month, I had better have a mammogram'" (Potter 2009d). In this way, distinctive packaging and marketing strategies have been used to build value for the brand, to make it socially and ethically calculable to consumers who are directly affected by and concerned with the cause.

In the fourth year of its association with the NBCF, CCA launched the Ribbon of Support campaign, which harnessed new social networking technology and, in an echo of the Well of Positivity, asked for messages of support for breast cancer patients to be tweeted. Participants were also invited to turn their Twitter avatars pink in support of the cause. While this particular campaign did not involve explicit donations from CCA, a company spokesperson explained that the "NBCF could never pay for the awareness that we raise" (Potter 2009d). In the CRM relationship, it is the brand itself that holds the most value for any philanthropic partner.

Since 2010, CCA has had a new "pink" partner in the McGrath Foundation, which funds "breast care" nurses throughout Australia. To mark this

Figure 8.3
"A message just for you," 2012.
Source: Coca-Cola Amatil/McGrath Foundation.

association, Mount Franklin turned its previously pale pink lids hot pink to reflect the McGrath Foundation's own branding. This aesthetic connection between the bottle and the cause reiterates the role of Mount Franklin in this relationship, to "draw attention' to the work of its partner and "encourage participation" (*Campaign Brief* 2012) in their campaign (see figure 8.4). In addition to its own "good work," CCA is consequently self-positioned as enabling a range of ethical activities by others: the cause, breast cancer public health awareness, is marketed by being attached to the sociotechnical capacity of the brand, and the brand is requalified as ethical by being attached to a well-known matter of concern. In this way the brand enables the social engineering of markets to become explicitly connected to difficult social issues, to become a participant in public health.[2]

In another campaign run over the 2007–2008 Australian summer, reforestation was targeted. This was a different cause, with another take on well-being and CSR. Through a partnership with the not-for-profit group Landcare, an organization reliant on corporate donations for the

Figure 8.4
Mount Franklin McGrath Foundation promotion, 2012.
Source: Coca-Cola Amatil/McGrath Foundation.

environmental remediation activities that it undertakes, CCA developed a Buy Me, Plant a Tree campaign. This short-term program received less exposure than the breast cancer initiatives, but still generated good public exposure. It raised $150,000, through which 250,000 trees were planted. Mount Franklin labels, rebranded with green borders around the logo for the period of the campaign, invited customers to register their bottle bar-codes online: in doing so, they would receive a tree planted in their name. From CCA's point of view, this strategy of personal association responded to consumers' desire to "help the environment": "We were just trying to enable our consumers," said a CCA spokesperson (Potter 2009d).

However, a backlash from within Landcare's own ranks, as well as from the wider public, forced CCA to defend itself against charges of "greenwash." A Landcare "senior official" publicly blasted the association as giving CCA "full power to 'almost rape the environment,' only for Landcare volunteers 'to put it back together'" (Lee 2008). In response, CCA called this view "the depths of ignorance." "We are committed to being a good corporate citizen," said CCA's director of media and public affairs in 2008, "particularly when it comes to water" (Lee 2008).[3] Much like the outrage expressed over the Brita water filters "drinking oil" marketing campaign, examined in chapter 7, the criticism garnered by CCA's Landcare association was a powerful example of the ways in which activism and issue networks interact with corporations, and specifically CRM. The activist accusation of greenwashing could be described as a form of "ethical counter auditing." According to Barry (2004, 207), this practice involves "expanding, interrogating and reframing the forms of ethical assessment commissioned and performed by the companies themselves." Bringing in other information about the environmental record and social impacts of companies allows challenges to the ethical concern and record of a corporation. This is not a reductionist critique based on a disinclination to acknowledge that corporations can be ethical. Rather, as the example of the Polaris Institute in chapter 7 showed, such an assessment often entails detailed and scrupulous research that expands the measures on which the corporation's ethical actions are evaluated. It also often involves a critical audit of the terms in which the corporation represents these actions.

The blog *Go Greener, Australia* undertook this kind of audit in response to the Plant a Tree campaign, setting Mount Franklin's claims of environmental efficiency in the production of its product against the calculations excluded by the corporation in reaching its figures: "Mount Franklin claims to be more efficient in their production and packaging than other manufacturers ... they don't mention the amount of oil used to make the plastic, though." "I'm all for planting trees," the blogger concludes, "but I don't think one tree per bottle is a good trade-off for the environmental problems caused by bottled water" (Grundy 2008). Ethical counterauditing is often driven by the desire to learn whether the execution of a company's stated commitment to ethics is merely marketing or a defensive protection of reputation. The critical intent of the audit is to ascertain whether activism will generate real effects and possible changes, or simply more strategic commercial calculations.

CCA reflexively responded to accusations of greenwashing and opportunistic association by repositioning the troubled Mount Franklin product in new networks of association and good works. As the single-use bottle of water came under increasing attacks by organized and disorganized campaigns, new uses and moral rationales for the product were sought. The idea of bottled water as an "essential service" (Potter 2011), stepping in where publicly provided water failed, became a prominent frame for CCA's public relations. As the following section explains, in 2007 CCA became publicly associated with a set of circumstances that saw this framing put to work in the highly charged context of a community health "emergency."

The Remote Communities Strategy (RCS) was a CCA initiative that targeted indigenous health concerns in remote communities in the Northern Territory, followed by Western Australia and Queensland. And this wasn't just a case of product association, of branding bottles with ethical attachments. Rather, the RCS targeted amenity and civic responsibility: it aimed to position CCA products as a legitimate solution to a public health crisis. This "innovative sales strategy," as CCA called it (Potter 2009e), offers important insights into the modern corporation's practice of biopolitics, and the ways in which commercial entities can leverage a political climate and policy program for their own purposes.

State of Emergency: Bottled Water as an Emergency Technology

In June 2007 the Australian federal government announced an "emergency response" to a commissioned report revealing disturbing levels of indigenous child abuse in the Northern Territory (Brough 2007). The then conservative coalition government announced an official "intervention" in direct response to these revelations. It imposed welfare regime change and heightened social policing measures across seventy-three indigenous— largely remote—communities in the Northern Territory.

Indigenous Australians are among the most disadvantaged social groups in the country, a complex legacy of colonial practices, policies, and attitudes. Some 30 percent of the Northern Territory's 231,300 residents are indigenous, the largest proportion of any state or territory in Australia

(Australian Bureau of Statistics 2011; Taylor and Bell 2001, 2).[4] Around 90 percent of these persons are classified as "remote"—that is, as living outside the greater area of the Northern Territory's capital, Darwin (Taylor and Bell 2001, 3). The relatively tiny size of Australia's indigenous population belies the vast history of social policy measures targeting this group, which have rendered indigenous people distinct subjects of biopolitical focus. Over time, such measures have encompassed an ongoing concern with managing the bodies, relationships, and monies of these people. Despite—or arguably because of—this attention, indigenous Australians have a significantly lower life expectancy than other Australians, as well as higher rates of key health conditions such as heart disease and diabetes.

The Northern Territory Emergency Response (NTER) in 2007 belongs to this history. While triggered by concern about sexual abuse, its rationale was quickly framed in terms of an indigenous people's welfare agenda. Its practices included the compulsory "quarantining" of welfare payments for food and other basic needs purchases, the government acquisition of control over townships to improve public facilities and housing, and changes to the management of remote community food stores to improve the freshness and range of items offered for sale. In effect, the emergency became identified as a crisis in community health. Ironically, to target *indigenous* well-being as an exclusive concern, the NTER had to suspend the stipulations of the long-standing Racial Discrimination Act of 1975, which prohibits the negative targeting of individuals on the basis of race. As a result, rights and freedoms previously granted to indigenous people, such as financial autonomy and the right to privacy (Gosford 2014), were withdrawn, excluding indigenous people from many prevailing definitions of citizenship.

The extraordinary nature of this suspension—an extreme measure in order "to protect children ... make communities safe [and] create a better future for Aboriginal communities in the Northern Territory" (Northern Territory Emergency Response Taskforce 2008)—was legitimated by the state's rationale to maximize the well-being of its people. Perhaps more critically, it was also legitimated by the concept of "emergency" that underpinned the NTER exercise. This was a striking echo of Agamben's (2005) theory of the "state of exception," according to which sovereignty is defined by the capacity to exclude subjects from the rule of law that grants to others certain rights and freedoms. The state of exception reframes sovereignty as not just a matter of power's exercise within a particular political order but as a more fundamental intervention into the life and death of political subjects. "The production of a biopolitical body is the original activity of sovereign power" (Agamben 1995, 6). Agamben's sovereign is the totalitarian state, a tyrannical force that, he argues, is increasingly indistinguishable from democratic governance (13). In the case of the NTER, however, the concern to administer indigenous life through the suspension of law saw the distribution of sovereign power to a host of nonstate actors, who also invested, for a range of interests, in the project of indigenous well-being. As Rabinow and Rose (2006) argue, the embrace of public-private partnerships in Western democracies in the last quarter of the twentieth century has meant that "sovereign power is no longer confined to those who are explicitly agents of the state—it apparently extends to those who have authority over aspects of human vital existence" (202).

In very similar terms, the NTER's umbrella focus on health offered a diverse terrain for new commercial interventions in matters of remote indigenous people's well-being. One key area was the long-standing issue of access to good-quality food and water for these communities—a concern clearly associated with high rates of obesity and diabetes, and early morbidity.[5] Community food stores are the single purchase point for fresh and packaged food stuffs in remote areas and have a notorious history of high prices and poor quality, driven by the difficulties involved in freight access to these isolated stores. The distance traveled by fruit and vegetables in particular means that items like apples can cost twice the amount paid for a can of soft drink (Gibson 2007). Preservative-heavy and precooked food are common staples, as are high-calorie beverages. A 2009 federal Inquiry into Community Food Stores, initiated by the NTER, heard submissions on this point. One stakeholder wrote, "As anyone who has worked in remote communities will tell you, sold food tends to be fried and salty, often pre-packaged and is accompanied by sweet drinks such as Coca-Cola" (Francis 2009).

CCA, the licensee of Coca-Cola in Australia, felt mounting pressure to act on this increasingly public image of its premium product being linked to indigenous people's poor health and disadvantage. A CCA spokesperson closely involved in the development of the RCS explained: "When I started at CCA several years ago, my boss said to me that I think we should do something in Indigenous health, I don't know what.... I can just see on the radar we're going to need to do something" (Potter 2009e). In a bad coincidence for the company, the publicity brought to these matters by the intervention was fueled by an announcement made on the Atlanta-based TCCC's website that the Northern Territory had "the highest per capita consumption rate of Coca-Cola in the world" (Gibson 2007). Understandably, this statistic gained widespread attention, in light of the obviously related rates of diabetes and obesity among indigenous communities in the Northern Territory. In 2008 an industry report revealed that Coca-Cola products ranked third, fourth, and fifth on the list of the top ten carbohydrate products purchased in remote community stores, after confectionary and sliced white bread (Glycemic Index Foundation 2009).

Globally, the association of soft drink consumption with poor health and obesity, particularly among children and lower socioeconomic communities, was already a prominent concern in public health discourse. According to the authors of *Liquid Candy*, a report on the health impacts of carbonated soft drink consumption in the United States, soft drinks were the single most consumed "food" in the U.S. diet, providing about 7 percent of daily calories for consumers (Jacobson 2005). The report's author identified aggressive marketing tactics by soft drink manufacturers as a key factor in these high rates of consumption, stating that in 2000, "the Coca-Cola Company, which accounts for 44 percent of the soft drink market in the US, spent over $200 million on television, magazine and other media advertising" (19). This kind of negative publicity had already promoted numerous campaigns in the United States to limit the company's use of exclusive contracts and sponsorship with schools.[6] The NTER and the Inquiry into Community Food Stores were shaping up to generate similar sorts of criticism and effects.

The brand threat posed by the associations between Coca-Cola consumption in indigenous communities and rates of ill health formed an emergency of its own for the company—one that demanded a swift PR response. Coke consumption was so ubiquitous in some communities that anecdotal reports circulated of mothers filling babies' bottles with the beverage in preference to milk. "From a brand perspective," related a CCA spokesperson who participated in developing the RCS, "these reports, whether true or not, weren't good" (Potter 2009e). The RCS was subsequently designed to maintain CCA's beverage dominance across Australia. It would prominently feature bottled water in a bid to educate its target consumers about the health benefits of drinking water. It would also promote bottled

water—specifically, CCA's leading bottled water product, Mount Franklin—
and wind back the visibility of its troubling high-sugar beverages, especially
Coca-Cola. The intention was never to replace sales of soft drinks with sales
of bottled water or other sugar-free beverages. Rather, it was to protect and
extend CCA's market range and profits. A company document puts the RCS
in these terms: "In 2006 our sales team in the Northern Territory came up
with an innovative strategy aimed at working with store-owners in remote
indigenous communities to increase the availability and take-up of spring
water and low-kilojoule beverages" (CCA 2011).

   Key to the RCS's work was the community store itself, which became
the object of a targeted sales and distribution plan that consolidated as the
NTER provided favorable conditions for CCA's commercial operations. In
2006, prior to the NTER rollout, the federal government had introduced
Outback Stores, a state-funded but independently operated network of
community food stores across the Northern Territory that was overseen by
a board of directors drawn from private industry, including Australia's biggest supermarket chains. Outback Stores received a funding boost with the
NTER, and its operational model proved favorable to CCA. A representative
of the company explained that Outback Stores brought "a very commercial
sense to its operations, which is probably why we found it so easy to work
with them"; "the government recognized that they're not good at running
supermarkets but there are other people who are" (Potter 2009e).

   The broad concern for indigenous people's health and positive nutritional outcomes that prompted the NTER's formation meant that CCA
could position its bottled water and other sugar-free beverages (diet soft
drinks and juice) prominently in community stores and complement the
Intervention's welfare objectives. Crucial to this strategy was the role of the
refrigerators that CCA supplied to remote community stores as part of its
RCS. CCA's coolers are a key part of the company's branding and product
placement strategies and are usually leased or given free of charge to stores
that sell CCA products. CCA's red Coca-Cola coolers are particularly iconic,
and thus, in a context that had turned against the high consumption of
this product, they became problematic to the company. Addressing this
issue, the RCS initiated a "Red to White" program to replace the red Coca-
Cola coolers with white Mount Franklin coolers. The significance of these
coolers was that they married product branding and presence with the provision of readily available, good-quality water (for a price) that was also

cold, in communities that commonly lacked access to both reliable potable water and electricity (see figure 8.5).[7]

The work of Mount Franklin bottled water in CCA's response to the attack on the brand was based on more than its status as a "low-joule" option. In keeping with the drink's new emergency use associations, the sale of bottled water in straitened indigenous communities offered a basic service—the provision of cooled drinking water—where public infrastructure was lacking. The absence of decent quality mains water is crucial to the story of indigenous health in remote communities. In a survey of 1,216 remote

Figure 8.5
Mount Franklin "white" refrigerator in a community store, Northern Territory, Australia.
Source: Coca-Cola Amatil.

communities in the Northern Territory, 784 communities were found to be getting drinking water from bores, ninety-nine from rivers or reservoirs, and fifty-one from wells and springs (Calma 2006). Groundwater across the region has often been polluted by mining and agricultural projects over many years. For example, in Utopia, a constellation of small remote indigenous communities 350 kilometers northeast of Alice Springs, bore water was discovered to contain dangerous levels of nitrate, and untreated water was being pumped directly into school drinking fountains (Skelton 2008). Hard bore water can also have bad effects on the kidneys, which has negative implications for those already suffering from diabetes (Calma 2006), while the foul taste and dubious quality of this water generally mean lower rates of water consumption in these communities.

The white Mount Franklin refrigerators offered an alternative infrastructure that could do the work of brand promotion, sales, *and* social welfare. They became a market device for encouraging consumers "to buy what they didn't normally buy" in the name of improving indigenous people's health. "With a fridge full of water, when people walk in and they can't see other drinks ... they go for water [instead]" (Potter 2009e). In 2009, CCA reported that these refrigerators had increased the visibility of the company's "health and well-being products" from less than 20 percent of retail space to over 40 percent (see figure 8.6). An internal promotional document produced by CCA cited the remote community of Hodgson Downs (Minyerri) in the northeastern part of the Northern Territory. Here the community store manager, Quin Hua, enthusiastically endorsed this change: "Thanks to CCA for coming all the way to install the new fridge in my store.... Three hours after it had been set up, I had sold thirty-three cases of healthy drinks.... I love those healthy drinks, [they are] better for me, and better for the community" (CCA 2009).

In an accompanying strategy, CCA removed all advertising of its high-sugar beverages from remote community stores. Instead, it introduced a new advertising campaign that, like the refrigerators, was presented as a kind of community public health service rather than a full-blown marketing strategy. The use of indigenous sports heroes to promote the consumption of bottled water was an important tactic in this campaign. As a CCA marketing representative recalled, "We've looked at how we can use those sporting codes to encourage people to think differently and to drink water"; "I had in mind something like a community service announcement, where

Figure 8.6
Mount Franklin products next to healthy vegetables in a remote community store, Northern Territory, Australia.
Source: Coca-Cola Amatil.

you'd have a sugar-sweetened soft drink and big bottle of water, and say, 'I drink this once a week, I drink this every day.'" The result was a 2008 poster series that featured retired AFL player Michael Long with the directive to "Choose Water" emblazoned above his head. His arm is outstretched, offering the viewer an unbranded bottle of water. According to CCA, the bottle is unbranded because "water sold in some of those communities is not always our brand … the idea is that you should just drink any water regardless of what it is" (Potter 2009e). However, in its unbranded everydayness, the bottle is normalized as the generic source of water, as basic infrastructure. While the poster encourages consumers to drink water, they are encouraged to purchase it too. Moreover, the bottle is recognizable in shape as a Pump bottle, CCA's bottled water product aimed at young men in the sports and fitness market (see figure 8.7).

Despite CCA's acknowledgment that "in those communities water is going to be the answer to a lot of the health problems" (Potter 2009e),

Figure 8.7
The Choose Water campaign, which ran between 2006 and 2010.
Source: Coca-Cola Amatil.

the RCS's emphasis on water consumption focused solely on container-ized water that was purchased, despite the well-publicized problems with municipal water access and low incomes in remote Northern Territory com-munities. What might be seen as a lack of public infrastructure was spun as an opportunity for market development: "In those communities that have bore water, it's not very good, so there's probably more of a place for packaged water" (Potter 2009e). Thus set to work as a commodity and a proxy infrastructure, bottled water in these remote community settings was promoted as a means of tackling community health crises, its advertising echoing the language of emergency brought to bear on remote indigenous communities by the NTER. Through a targeted sales strategy, CCA's bottled water products became a way for the company to promote itself by appearing to support a marginalized and disadvantaged population while stepping in as a nongovernmental actor to supplement an inadequate public system of basic service provision.

Conclusion

Ethical branding and CRM and CSR programs such as the Remote Communities Strategy are different ways in which beverage companies have sought to "spin" the bottle. If ethical capitalism involves diverse strategies for rendering markets responsive to controversy, for reconfiguring them in relation to their explicit and implicit political effects, then the examples looked at in this chapter are powerful evidence of this development. They are also powerful evidence of the need to examine such practices in their specificity, for each involved very distinct strategies for experimenting with market assemblages and requalifying bottled water as an agent for good. While each instance shows how markets can reconfigure the relations between and among economics, politics, health, and the environment, the ways in which such reconfiguration is practically organized and the devices that are used to do so are immensely variable, ranging from personal narratives to pink lids, Twitter posts, and indigenous footballers' endorsements. More critical is the need to assess the impacts of such developments, to ascertain the ways in which they reflect effective responses to the wider contestation of bottled water as a very problematic market thing. Do these initiatives represent significant new market *agencements* in which diverse actors actively reconfigure bottled water in the interest of generating better social and political effects? Or are they evidence of antipolitical strategies designed to contain and incorporate threats and defend commercial calculations at all costs?

Each example examined here produces different answers to these questions. In relation to ethical branding, it is possible to argue that this market arrangement seeks to make explicit many of the problematizations of bottled water as it participates in the wider biopolitics of water. In devising marketing themes around thirst and the absence of effective water supply in many places in the global south, these niche markets connect water politics with distinct market arrangements. Rather than purify bottled water markets by claiming they have no relation to thirst and water scarcity, One Water and related examples cast the product as a participant in strategies to address these problems. The brand and its consumption are wrapped in philanthropic gauze, and the market is explicitly configured to combine economics and politics (Callon 2009, 542; see also Varadarajan and Menon, 1988).

CRM is another case altogether. As we saw with Mount Franklin and CCA, marketing around a cause generally emerges in large multinational corporations with established brands. These complex and extensive market arrangements are essential to promoting the cause effectively by connecting it to existing networks of recognition and brand reputation. What is interesting is the nature of the causes that are selected to market, as a way to generate ethical qualities for the product. Unlike in ethical branding, the selection of these causes is usually disconnected from any explicit political contestation of the market or anti–bottled water activism. Plastic pollution, campaigns to defend public water supply, and the like were not chosen by Mount Franklin as appropriate causes; instead, it was breast cancer—an issue that initially seemed to be utterly unrelated to bottled water markets. (A similar point can be made about the lack of any obvious connection between tree planting and disaster relief.) Attaching the bottle to these causes and marketing its role in relation to them were devices designed to amplify the appeal of the product to its primary consumers—no doubt based on extensive market research. They were also measures designed to demonstrate the corporation's commitment to causes and its ethical awareness to a wider audience. However, as Barry (2004, 204) so convincingly writes, "The success of a demonstration depends, in part, on whether it is not viewed by its audience as simply a manifestation of the political and economic interests of the demonstrators." It is impossible to ascertain whether consumers viewed these demonstrations as altruistic displays of generosity and concern for the public good or as part of the corporation's general pursuit of its economic interests, part of its ongoing work in managing and exploiting the brand. In an era in which brands touch the most intimate relations of everyday life, connecting them to annual mammograms seems utterly normalized.

What is more important is the way in which marketing causes emerge as part of the reflexive reconfiguration of beverage market assemblages. This practice is inevitably implicated in the product—the bottle of water—becoming a political object. This emergence has the force and capacity to make trouble for the market frame. Attaching the bottle to virtuous causes through CRM implicitly *detaches* it from controversy, from difficult causes that emerge directly from the effects of the market. It is a device designed to be antipolitical. In the final case we looked at, the issue of emergency and biopolitics pointed to a very different register of effects, one in which the corporation sought to become a substitute for safe public infrastructure and to be normalized as the provider of potable water. In this instance, bottled water was employed by the corporation as an instrument of moral and technical governmentality; the pressing issue of indigenous people's disadvantage and ill health in remote communities was addressed through a marketing strategy that sat within a set of state and nonstate responses to a declared emergency.

In relation to water scarcity and a growing diabetes and obesity epidemic, bottled water participates as both a problem and a solution: the effects of opportunistic bottled water markets in places where safe public supply is not established undermine the struggle for investment in infrastructure, while the health problems caused by excessive consumption of sweetened beverages are "solved" by switching to water. However, by claiming that "water is water," bottled water producers deny these contradictory effects. CCA's activation of bottled water as a legitimate and sustainable source of potable water in remote indigenous communities performed this equalization. The serious problem of inadequate public infrastructure *and* the political event of a brand under attack were addressed by a sales strategy that ultimately saw sales of bottled water *and* sugar-heavy soft drinks continue to rise in the Northern Territory[8] (Brimblecombe and Thomas 2010; Potter 2011). Responding to a market threat meant embracing rather than obscuring the biopolitical nature of bottled water; it meant incorporating concerns over the public availability of water into an evolving rationale for drinking bottled water. As One Water Australia now also turns its attention to achieving "water security" for impoverished remote indigenous communities (One Water Australia 2013), this version of ethical capitalism employs a strategy markedly similar to its corporate counterparts. At the heart of each, despite their distinctive modes of reflexive organization, is the suppression of the real material, biophysical, and social distinctions that produce different waters—that is, water as a common good and water as a product.

9 Conclusions

In March 2014, San Francisco's Board of Supervisors voted to ban bottled water sales on city-owned property. Violators found selling water in twenty-one-ounce single-serve or smaller bottles would be liable to fines of up to $1,000. The reasons given for this ruling, which was reported around the world, were protecting the environment, combating climate change, and reducing plastic consumption. The local newspaper, the *San Francisco Examiner*, reported that the beverage industry had "predictably protested" (Timm 2014). A month later the public water utility, DC Water, was awarded the Grand Prize in the Environmental Communications Award from the American Academy of Environmental Engineers and Scientists for its Drink Tap campaign. The challenge DC Water set itself in this social marketing exercise was explained like this:

Americans have become less connected to their drinking water systems, turning to bottled water out of convenience, due to water quality concerns or in response to successful marketing strategies from a $12 billion U.S. bottled water industry. The goals of the Drink Tap campaign are to restore consumer confidence in drinking water systems, educate the public about the benefits of tap water and to increase consumer preference for tap water. (American Academy of Environmental Engineers and Scientists 2014)

These are just a few of the many recent examples we could cite to show how state authorities, from city governments to water utilities, are fighting back against the bottle through bans, publicity campaigns, the reintroduction of drinking fountains, and similar efforts. Equally notable is the rise of branding strategies for public utilities that emphasize civic rather than commercial values for water. Here, the state is distinguished from markets and the economy in the interest of strategically demarcating its activities as outside the sphere of commercial interests and calculations. This demarcation is not evidence of a clear boundary but a deliberate political technique (Barry 2013, 41). In all these cases, bottled water has been both problematized and generative. The creation of an economy of qualities for tap water becomes meaningful only in relation to bottles; the framings and effects of bottled water markets function as a surface or reference point from which new governmental practices and valuations emerge and find justification.

While these state responses to bottled water markets are important, our argument throughout the book has been that they are one of many forms of contestation that have emerged around the bottle. Although the techniques deployed reflect the institutional rationalities and devices of government, their aim is not simply to oppose but to rearrange various aspects of existing water provision in order to revalue the ordinary properties of tap water. Another key effect is to charge the mundane bottle of water with great moral import. Obviously, these strategies are meaningful only in places where tap water provision is safe and reliable. Where it isn't, bottles are often perceived and used as the potable water infrastructure, as the source of trusted supply. In these settings, the bottle is an essential element in complex and messy networks of water provision. These examples of the different ontological status of bottled water show that the capabilities and effects of things cannot be taken as given. They also show that the politics of bottled water are far from self-evident: they have to be enacted, and in these processes the character and ontological status of the bottle and the water change. Political analysis cannot be reduced to generalized oppositions such as tap water versus bottled water, or water as a public good versus water as a private good. Instead, it is essential to understand how political situations emerge, the grounds on which bottled water becomes problematic and morally troubling, and the specific devices and artifacts that are drawn on to contest its normalization.

Another key argument we have developed is that markets, like political situations, have to be assembled. Rather than focus on large macroprocesses such as corporatization or neoliberalism, we have turned our attention to the mundane devices, discourses, and calculative techniques that have made bottled water markets possible. Central to our analysis has been a concern with the material and technical capacities of the bottle. Although packaging remains largely underappreciated in many accounts of markets and consumption, in the case of bottled water, it is impossible to avoid. As we showed in chapter 1, the invention of the PET bottle was essential to turning water into a fast-moving consumer good (FMCG). The question was how the bottle's technical capacity became implicated in rendering water calculable. How did this mundane thing interact with water to transform its qualities and the everyday practices of drinking? In pursuing these questions, we have not claimed that everything can be reduced to or is determined by the bottle; rather, the technical capacities of PET are contingent on and emerged in relation to a range of other market devices and techniques. And when it came to packaging water, these technical capacities were enhanced and elaborated, and came to have powerful transformative effects.

In his analysis of the introduction of canned foods into American retailing in the 1920s, Franck Cochoy (2013) argues that this new form of containerization did more than simply transcend seasonality and preserve food more effectively. As canned goods were introduced into shops and consumers began choosing them, the technical capacities of the tin can became powerful market drivers. Their stackability meant they could get off the shelf and be displayed on shop floors or counters, and their opacity meant that shoppers had to pick them up and read the labels and brands to learn about the product, thereby bypassing the mediations of the vendor. These developments implicated the technical capacities of the can in shifting consumer practices and "clearly prepared the move of the grocery store to self-service and mass consumption" (Cochoy 2013). This argument has parallels with our analysis of the PET bottle. Like Cochoy, we have sought to understand how the bottle's sociotechnical capacities were expressed in beverage markets and how its practical packaging functions generated multiple effects and impacts that were far from exclusively "economic" and far from self-evident. There is no question that the mundane plastic bottle has been a leading market driver in the growth of water as an FMCG. In contrast to the opacity of tin cans, the bottle's transparency seemed to enhance the purity of the water, enabling consumers to feel that they were directly accessing a liquid far superior to the unbranded flows of the tap. Its lightweight and inviting retail refrigerator displays also suggested not just self-service but portability. The resealable lid and graspability prepared consumers for the rise of new discourses of hydration and the habit of carrying water everywhere so as to be able to constantly sip. These are a few of the many ways in which the mundane bottle supported the advent of significant shifts in the distribution, mass retailing and consumption of water. The point is not that packaging water was inherently bad but rather that in the assemblage of water as an FMCG, packaging, in relation with various other market devices, acquired potent forms of agency that reverberated far beyond the internal organization of markets. And it is these reverberations that have constituted the ongoing event of bottled water evident in the actualization of new relations between water, container technologies, and mass consumption.

In assessing the event of bottled water, we have chosen to scrutinize patterns of emergent causation rather than the logic of inexorable market expansion and corporate intentionality. This decision helped us shift our thinking from reductionism to relationism, and to an analysis of the ways in which processes of market emergence have generated a multiplicity of effects. Our concern has been to trace some of these effects without recourse to the critical assessments and judgments already out there. It is not that we dismiss these assessments; rather, we have sought to create other ones, to get beyond imposed frameworks and pay close attention to how the bottle's capacities are enacted in different settings. In many existing critiques, bottled water is assessed as threatening the fundamental biopolitical value of water. In this approach, the biopolitical arena is often rendered in somewhat romantic and transhistorical terms. Water is recognized as constitutive of biological and social being, and any threat to it from bottles and capital becomes a threat to life and basic human rights. While we understand this activist strategy of linking the specific to the universal, we have sought to challenge it. Rather than see markets in bottled water as a threat to a fundamental water biopolitics, we have argued they represent the emergence of new forms of calculability for water that are not simply economic but also ethical, affective and variously political. Drinking water from bottles constitutes very distinct biopolitical regimes that are a product of the history and context of markets and their interactions with other forms of water supply. In this way, then, bottles don't destroy but invent new forms of water biopolitics.

However, despite the immense sociomaterial and cultural variability in how bottled water markets and drinking regimes are organized, it is possible to identify three powerful emergent effects when water is packaged and commodified in single-serve plastic bottles. These are the capacity of the bottle to discipline water's biophysical unruliness, the ways in which bottles and brands appear to manage water risk and scarcity, and the individualization of water provision. In 2003, Karen Bakker influentially described water as an "uncooperative commodity." Although she does not consider bottled water in her far-reaching analysis of global water politics, her work has been invaluable, and by way of conclusion, we situate our analysis of bottled water's emergent political effects in relation to some of her key claims. This strategy allows us to understand the rise of the bottle not only in relation to wider debates about the biopolitics and governance of water but also in relation to possible water futures.

For Bakker, full commodification of water is notoriously difficult. She argues that water is a natural resource that poses significant barriers to the conversion of its use value into exchange value, and therefore resists being completely appropriated. The key reasons why, for Bakker, water is "uncooperative" are, first, that it is a moving material or flow resource, and therefore not easily bounded. This makes the establishment of full property rights difficult because it is unclear where the boundaries of the property are. It also means that users upstream can affect the water downstream, so private ownership of the location of a water source does not guarantee control over its quality. Private property regimes cannot manage negative territorial effects. And second, water is heavy. Its density makes it difficult to move around, so that water is "expensive to transport relative to value per unit volume" (Bakker 2003, 33). Distributing water to populations therefore mandates expensive infrastructure—something that prevents many companies from entering the market at the level of infrastructure development. This reality also favors local distributive networks to reduce the cost of movement. Water is most cooperative when it doesn't have to be transported too far. These are the main aspects of water's various biophysical realities that Bakker identifies as uncooperative or recalcitrant. Not only do they challenge the idea of water as the passive object of exploitation and capital accumulation, they also show that, in many ways, these realities implicitly favor some form of state participation in distribution as the most efficient and equitable method for connecting water with populations. This doesn't mean that full commodification of water is not possible. Rather, these characteristics of water set the terms with which any provisional arrangement for water marketization must contend.

For these reasons, water can be understood as provoking very distinct market forms able to discipline its unruly and unpredictable materiality.

As Bakker (2003, 32) argues, when water becomes involved in processes of marketization, its unique biophysical properties both enable and constrain its own production. Markets work by organizing a set of relations, technical devices, and calculations that aim to discipline and frame this biophysicality and make it amenable to exchange. In this sense, they are relational enactments of the material or biophysical qualities of water. And in this enactment, water participates as an actant, effecting certain actions according to the way in which its biophysical affordances submit to or resist market relations. This is where bottles appear to have a distinct technical advantage and to have become a potent force. Bottles offer a way around many of these biophysical, technical, and legal constraints. They discipline and singularize water through the logic of containerization, most notably the single serve. While beverage companies have to negotiate these broader legal and technical realities at the source of collection and production—and doing so often involves ruthless forms of primitive accumulation of water sources—in the moment of delivery to the consumer, the bottle appears as a mobile, alienable, and personal water source. As we argued in chapter 1, the container appears to become the source of what it has stored and preserved: it is a device for reframing and re-sourcing water.

The second generic effect of bottled water markets is to appear to manage risk and scarcity. These terms have become central to the administration of water over the past thirty years. As Bakker notes, the rise of discourses about state failure and the material evidence of water supplies under threat have often created the conditions for markets to be promoted as the ultimate fix. During this period, market management and its inherent "efficiencies" were often linked to better service provision, and the care and conservation of water. Bakker terms this alliance between the logic of the market and the logic of proper environmental management "market environmentalism"—the entanglement of diverse economic strategies with nature, or the increasing use of market forces and capital accumulation to address growing environmental issues (Bakker 2003, 26–28). The key characteristics of market environmentalism with regard to water were produced through a reorganization of the relationships between and among water, users, valuation measures, and administrative practices. The political rationality of securing supply that had driven water infrastructures since the nineteenth century was replaced by a concern with the threat of water scarcity, contamination risks, and the need to regulate demand. In this shift, water became requalified as a vulnerable and scarce resource, and techniques of conservation and the search for alternative sources—from recycling to household reduction campaigns—dominated new forms of provision. Pricing also shifted to techniques more attuned to measuring actual volumes used rather than a service fee for unmetered public access. This change was justified on the basis of making users—who were now "customers"—more aware of the need for careful and sustainable water practices. In short, scarcity generated pressure to consider water as a form of property—or at least a good that was amenable to measurement, calculation, and careful allocation.

Bakker (2009) avoids describing these changes under the blanket term of neoliberalism, for this term often obscures the specificity of the reforms that were enacted during this period, and the various types of markets they produced. Privatization, marketization, deregulation, commercialization, and corporatization of water provision, for example, involve very particular political and sociotechnical conditions and have had different impacts on water resource management, governance, pricing, institutional players, and households (Bakker 2007, 436).

These reforms may be interconnected in the sense that one may prefigure or lead to another, but they are also specific assemblages. What they reveal are the unprecedented shifts in the political and economic organization of water utilities that have occurred over the past thirty years and the effects of these shifts on wider social and everyday perceptions of water quality and supply. The rise of discourses of scarcity and risk, prompted by droughts, contamination scares, and dramatic price increases, signaled the end of unlimited cheap supply—the end of never-ending flows. And it is in this context that the bottle of water emerged as a potent alternative. As supermarket shelves, refrigerators, and vending machines full of thousands of bottles began proliferating, the commodity form—branded, purified, and chilled —came to represent a zone of abundance. The seriality of the bottle, its sense of endless massification, seemed to provide an alternative to the anxieties generated by a diminishing or contaminated public supply. There is no doubt that, for beverage companies, the rise of discourses of water scarcity and risk was seen as a significant market opportunity.

Finally, the bottle offers the opportunity to individualize provision, to make water into a discrete private good and a form of portable property. When a utility provides water that is unbranded and from an unknown source, the user consumes it on the basis of trust, on the basis of a belief in the responsibility of the utility to deliver a mass service and a public good that is safe and reliable. Bottles reorganize this network of relations and economy of qualities; they reconfigure responsibility and reliability into a contractual exchange between the individual consumer and a beverage company. The bottle remediates water through the semiotics of the brand, technical information about its source, its biochemical purity, and role in personal health management. It implicitly undermines the accountability of public or quasi-public provision by differentiating the water contained in the bottle from the tap. Unlike water from a tap, water in a bottle is attributed a biovalue and a clearly established provenance. The bottle, then, is a technology of detachment, a means to avoid the undifferentiated water service and choose the branded product. In this way, responsibility for safe drinking water is transferred to the individual, to the choices individuals make and the ways in which they perceive the branded bottle as delivering "better" water.

While it may seem that these three key effects reflect more fundamental structural oppositions between taps and bottles, citizens and consumers, or public versus market forms of calculation, we have argued that bottled water markets often generate forms of "clandestine hybridization" between these categories—that diverse political situations emerge in processes of contestation and creative contamination with markets. The tap versus bottle opposition, for example, is not the self-evident sum of the politics of bottled water; it is one of several tactical and highly situated political events in which a market and its externalities are strategically linked to wider state processes and water governance. Barry (2013, 84) calls this process of linking a specific dispute to a larger political situation "abduction." Abduction points to causal agency and infers antecedents from consequences. "It both turns audiences towards and constitutes the existence of forces beyond the object or event itself" (2013, 84). Abduction isn't reductionism it is a process of inference that becomes performative. And, as we saw in chapter 7, it has been a central practice in the formation of the diverse issue networks and activism that have emerged around bottled water. Abduction has also been a central practice in the development of ethically branded bottled water markets—such as Thirst Aid Water, explored in chapter 8. There the value of branded water was generated through explicit links to political situations, often operating at vast distances from the particular geographic reach of the market. The clandestine hybridization at work in this example involved connecting existing global water crises and development debates to bottled water market frames in such a way that consumers could justify their choice of drinking water from bottles as moral rather than self-interested.

However, overt controversies, the emergence of issue networks, and ethically branded waters are by no means the totality of political situations that exist in relation to bottled water. We have also traced the ontological politics that surround the rapid growth of markets in packaged water. Ontological politics often work in the shadowy registers of "interference" with existing realities. Realities are being practiced everywhere: they are complex, uncertain, and interact with each other to produce difference (Law 2004a, 67). And it is in these differences that ontological realities can become ontological politics because difference can generate both conflict and dissent or the imagination and enactment of alternative realities. In part II we investigated how bottles came to matter in three Asian cities, how diverse packaged water markets, drinking, and disposal practices emerged and were made meaningful in relation to existing arrangements. While we were concerned to trace how bottled water was normalized in a range of settings, we were also interested in the effects of the "becoming ordinary" of bottled water—of how new bottle practices interacted and implicitly interfered with the broader struggle for public access to safe potable water.

Although many bottled water marketing campaigns emphasize the technical benefits of water delivered in bottles—convenient, portable, cold— thus framing the bottle as a handy resource when other forms of supply are not available, these framings never exist in isolation. They are usually part of a repertoire of claims that, as we saw in chapter 2, implicitly interrogates nonmarket forms of worth for water. As William Davies (2013, 36) has argued, "the success of market rationality does not simply depend on its capacity to 'perform' in a technical sense ... but [also] in a public and rhetorical sense, where it must seek to resist other forms of rationality, or else colonize or negotiate with them." Our claim in this book is that the assemblage of bottled water markets has involved—though in highly differentiated and situated ways—numerous forms of negotiation with nonmarket forms of worth for water. These negotiations have often been exceptionally effective in undermining appeals to the public or common good, and in reconfiguring the value of water in terms of personal freedoms, self-care, or self-interest. This is the event of bottled water, the ways in which markets have acquired the capacity to reframe drinking water as an individual practice rather than a common or shared experience, the ways in which the collusion between PET bottles and water has inaugurated new regimes of drinking.

To bottle water, to brand it as a beverage, and to deliver it in a single-serve disposable PET bottle is to interrogate and reorder biopolitical relations in multiple ways. We have shown how bottles and drinkers and water form associations endowed with distinct biopolitical implications—how the relations between these participants generate "regimes of living," or particular configurations of normative, technical, and political elements that render "living" subject to ethical interrogation and query. This effort involved close attention to the agency and alliances of PET and water in these regimes, in order to understand exactly how they acquire ethicopolitical capacities—how PET bottles become participants in assembling new relations between water and life that do not automatically overrule other water regimes but interact with them in complex and often very troubling ways. The challenge was to assess how interaction becomes interference, how bottled water realities acquire the force to displace or disrupt other more equitable and sustainable ones, how commodity form becomes destructive biopolitical capacity.

Despite the immense sociomaterial, geographic, and cultural variability that we have documented in this book, bottles generate recurrent biopolitical effects. These effects play out in different ways with different emphases in different places, but the alliances bottles form with water are powerful and repeated. These alliances relate to the capacity of the bottle to discipline water's biophysical unruliness, to appear to manage risk and scarcity, and to individualize and brand water provision. What happens to water in these dynamics is that its capacity as a basic material of life is transformed. The PET bottle makes water into a market thing, an object of consumption. In this transformation, bottles repress water's universal role as a crucial participant in making a just and more-than-human common world. This is the destructive biopolitical capacity of the PET bottle. In the face of our empirical investigations of how bottled water markets, consumers, and politics are practical and contingent achievements, a persistent bigger question has also been in play: how to deliver drinking water in just and fair ways. A core thesis of this book is that the bottling of water, unlike the bottling of orange juice or Coca-Cola, troubles assumptions about the universal function of this vital liquid in enabling all life. A potent and unexpected effect of the rapid growth of water as an FMCG has been the amplification of water's critical role in sociality and in symbolically delimiting spaces for the enactment of the common good, free from economic imperatives. This is not to argue that the common good can only be established through the disavowal of economic processes, but rather, to realize how markets have enabled new forms of calculability that put into sharp focus the question of what the limits of economic practices and forms of worth should be when it comes to water. Should water be classified as too special to be exposed to market-based forms of valuation? Should it be regarded as an exception because of its fundamental implication in the life of the commons? The emergence and spread of markets in bottled water over the past quarter century reveal that this is far from the case: there are no guarantees. But this is what makes bottled water such a compelling and volatile political object— something that is not only materially but also philosophically uncooperative as a commodity. However it is articulated, in whatever arrangements or forms, bottled water continually provokes normative questions about the sustenance of life and the common good.

Notes

Chapter 1: Packaging Water

1. The recent emergence of markets in plastic bottled water does not deny the historical reality of already existing markets in bottled water, but such markets were largely niche or prestige not mass and the water was most often contained in glass.

2. "Unfinished objects" is Knorr Cetina's (1997) term. She uses it to explain how objects are central to sociality, how they have an open and unfolding character, and how they lack "objectivity," or a fixed and unchanging identity.

3. The idea of the actual and the processual draws from A. N. Whitehead's (1978) elaboration of process philosophy.

4. See DeLanda (1995) for an account of the expressivity of materials. This is akin to Bennett's (2010) idea of "vibrant matter" in the way that it acknowledges materials as always capable of exceeding their social determinations.

5. See Meikle (1995) for an account of the impact of the squeezable ketchup bottle.

6. Thanks to Andrea Westermann for the wonderful term "bottleability."

7. My use of "emergence" here is shaped by Harman's (2009) discussion of Latour in *Prince of Networks*. Harman argues that Latour's position is that things are defined by their relations and their outward relational effects on other things, not by their internal composition: "Latour veers toward a functional concept of emergence: a thing emerges as a real thing when it has new effects on the outside world, not because of any integral emergent reality in the thing itself" (Harman 2009, 158).

8. See "PET Remains a Favourite" (2008).

Chapter 2: From Containers to Contents

1. This is Wilk's (2006) argument. Despite an incisive analysis of the rise of bottled water, his account of branding focuses more on the representational logics of bottled water brands than on their complex performativity as market devices.

2. We capitalize Evian in this analysis rather than use evian, the current brand orthography. The use of the lowercase was introduced in the 1990s in order to detach the brand from the city and give it more "flexibility" (Evian 2011). Most of our analysis in this chapter focuses on the period before this change—hence the use of uppercase.

3. See Porter (1990) for an account of the history of water treatments.

4. See Michael (2000, 26) for an excellent discussion of Michel Serres's notion of the quasi-object.

5. This origin story is outlined in the Évian tourist guide, available from the information center. It also drives the exhibition on the evolution of the water and the brand at the bottling factory on the edge of town. See also Green and Green (1985).

6. The water from the Cachat spring was analyzed in 1807 and its biochemical composition and unique mineral concentrations were identified. They have been found to be consistent ever since (Dana 2000).

7. Only fifty water sources in France have achieved this highest classification; others are classified with a lesser appellation, *les eaux de source* (Green and Green 1985).

8. This history is set out in the Evian Museum at the bottling plant.

9. See Lury (2009, 74) for a discussion of "the mass" as a heterogeneous population open to multiple configurations, such as tribes, fans, and lifestyles.

10. See Hein (2002) for an account of this marketing strategy.

11. See Clarke (2007) for an account of these attacks on the public water supply.

Chapter 3: Frequent Sipping

1. A version of this ad appears at http://www.evian.com/en_INT/54-evian-PET-bottles (accessed May 12, 2012).

2. The revival of hydrotherapy in Europe in the eighteenth and nineteenth centuries consisted of various therapeutic practices, including hot and cold baths, drinking cures, purging, and mud treatment. The bottling of water in this context served as an attempt to capture and transport the reputedly therapeutic properties of the water found at renowned spa destinations and resorts. See, generally, van Tubergen and van der Linden (2002).

3. The information on hydration discussed here is accompanied by a special offer, made to private health insurance members, concerning the annual rental of water coolers. Water is easily transformed, in proximity to the discourse of hydration, from a public to a private service. On the blurry distinction between good and service and its significance for practices of water provision, see chapter 4.

4. Indeed, with the rise of "ethical consumption" and ethically branded bottled water, the act of purchase may itself be conceived of as a form of social participation in which private acts of purchase are thought to discharge the moral responsibilities of citizenship. See chapter 8.

5. As Isabelle Stengers (2008) argues, scientific experiments do not uncover an objective, stable world. Rather, they emerge from processes of "mattering," which are always situated and motivated in particular ways.

6. The conceptual shift from statements to propositions attempts to register the multiple articulations brought about by scientific experiment, rather than resting on a fixed sense of primary qualities. As Latour (2004, 213) puts it, "The universe is made of essences, the multiverse … is made of habits."

7. The interest in running and aerobic activity was driven in part by cultural apprehensions around bodily inactivity. The shift from manual to white-collar work after World War II sparked middle-class concerns about the sedentary nature of office work. During the 1960s, a series of epidemiological studies had raised public awareness of chronic diseases such as heart disease and cancer, with the surgeon general's report on smoking appearing in 1964. The focus on risk, environmental hazards, consumption, and the workplace created new associations in the public imagination between lifestyle and disease, prompting new investments in "lifestyle change" as a means to achieve better health. Meanwhile, the increased leisure time among the middle classes enabled people to pursue these activities as a form of recreation. See generally Crawford (2006).

8. Here we draw from Lauren Berlant's formulation of "cruel optimism"—a phrase she uses to describe certain similar attachments when they come to operate as an "obstacle to your flourishing" (Berlant 2011, 1). We hesitate to use the phrase "cruel optimism" itself, as the criteria by which an attachment might be judged to constitute an "obstacle to [one's] flourishing" are not entirely clear in this instance.

9. "Biovalue" is a term originally coined by Catherine Waldby (2000, 3), who considers it to be "generated wherever the generative and transformative productivity of living entities can be instrumentalized along lines which make them useful for human projects." While Waldby seems to be referring primarily to research and development of organic tissues, the concept's association with processes of capitalization and surplus value makes it an apt descriptor for modes of value-creation that converge on a whole range of goods and resources associated with health in the context of capitalized medicine.

10. This aspect illustrates how hydration discourse has come to incorporate contemporary concerns around obesity and diabetes. In the hands of the beverage industry, hydration becomes a hybrid form of health messaging, capable of absorbing many different health concerns and medical etiologies.

11. There is no "sedentary" profile. Some may feel miffed, and others relieved, to find that the choice for a middle-aged person who is not pregnant is between Over-Indulgent and Active.

12. Interestingly, most of the expert commentary contained in these materials—from the H4H Meeting expert testimonial videos to the medical references listed on the "educational slide kits"—pertains to concerns about sugar and obesity, rather than "hydration," as originally defined. The concern with the aerobic body, which initially propelled studies of hydration, has largely been replaced by a discourse that poses water as the healthy alternative to sugar-sweetened beverages—perhaps more justifiable or at least in line with contemporary medical evidence. Here, water is positioned as "the only liquid your body needs to hydrate," and therefore as essential in "maintaining your caloric balance."

Chapter 4: Drinking Water Arrangements in Bangkok

1. See Anderson (2012, 6) for an excellent discussion of the hazards of essentializing "Asia."

Chapter 5: Enacting Water Scarcity in Chennai

1. Some 1,200–1,300 millimeters, compared with 800 millimeters per annum.

Chapter 6: Practices of Economization

1. Research for this chapter was conducted during two visits to Hanoi's plastic villages in 2010 and 2014. Access to the villages and information about plastics solid waste in Hanoi was facilitated by Pham Van Duc and Nguyen Thi Le Huyen from URENCO; translation and extensive support and assistance in the villages were provided by Le Huy Tuyen and Dr. Minh Duong; expert technical research and assistance were provided by Warwick Pearse. We thank them all immensely. We also thank the anonymous plastic bottle recyclers who allowed us to watch them work and ask numerous questions.

2. There has been considerable discussion about the waste effects of bottled water and of the global accumulation of plastics waste and pollution more generally. In relation to bottled water, Clarke (2007) gives a detailed account of the poor recycling record of plastic water bottles despite industry claims that bottles are recyclable, and therefore "environmentally friendly." He shows how beverage corporations in the United States have vigorously fought the introduction of container deposit legislation and other "bottle bills" that would tax the industry for the costs of managing packaging waste. Numerous other reports have documented the phenomenal increase in plastic bottle waste in urban waste streams, paralleling the growth of the

bottled water industry (see Container Recycling Institute 2003). MacBride (2011, chap. 5) provides an excellent account of the scale and complexities of plastic waste as the definitive "modern waste": synthetic, unpredictable, and heterogeneous—and therefore difficult to recycle. It also poses significant problems to health and ecosystems, and is often exported to the developing world for processing under very uncertain conditions (183). In relation to the accumulation of plastics waste more generally, there is a massive literature within many branches of environmental science on the global plastic waste crisis (see Gabrys 2013 for a discussion of plastics in oceans and Barnes et al. 2009 for a discussion of the accumulation of plastic fragments and debris in global environments; see also Ellwood 2008; Cormier 2008; Grant 2009).

3. See Gill (2012) for an account of a similar process in India.

4. See, for example, DiGregorio (1994), Gill (2012), and Godden (2011).

5. See Gibson-Graham (2006) for an account of the performative dimensions of economic subjectivity.

6. This point was made in an interview with Pham Van Duc, deputy general director, URENCO, April 2, 2014.

7. Interview with Pham Van Duc, deputy general director, URENCO, April 2, 2014.

8. See MacBride (2011, 183) for an excellent discussion of the problem of the heterogeneity of plastic waste and the challenges this presents for effective recycling.

Chapter 7: Contesting Bottled Water Markets

1. It is impossible to cite all the anti–bottled water campaigns over the last nine years, as there have been so many in a variety of forms. The following list offers a small sample: Angel (2008), Corporate Accountability International (n.d.), Hickman (2009), Olson et al. (1999), and Story of Stuff Project (2010).

2. See Bakker (2010) for an extensive discussion of these and other antiprivatization of water campaigns.

3. See Callon, Lascoumes, and Barthe (2009, 237) for a discussion of the problems of representation in hybrid forums.

4. See Foster (2008, chap. 6) for an excellent analysis of the campaigns against Coca-Cola.

5. Information on the FilterForGood website about the environmental impacts of bottled water has parallels with the use of health information about hydration needs outlined in chapter 3. Information ostensibly offered in the name of public health is also, at the same time, creating a consumer.

Chapter 8: Spinning the Bottle

1. Cause-related marketing (CRM), as the name implies, is a clear marketing strategy that works through processes of branding and product promotion to "support" certain ethical causes. These causes are "attached" to particular products, and thus associated with the company behind them. Corporate social responsibility (CSR), on the other hand, refers to a broader, multifaceted set of strategies that aim to prove the ethical and socially responsible nature of a corporation and in doing so to associate its products with good work. Such strategies aim to establish the ethical credentials of an entire organization. CRM often operates as part of a CSR program.

2. Mount Franklin's decision to associate itself so strongly with breast cancer research, treatment and "support" sits in a further context of links drawn by activists between the relationship between bisphenol A (BPA) in plastic and endocrine disruption in the United States and other countries. This activism is not extensive in Australia, however.

3. Mount Franklin's Pink Lids campaign has also been accused of exploiting a cause in order to redeem the product. The charge of "pinkwashing" is widely employed by critics of the almost ubiquitous association of many corporations and products with the breast cancer cause. As one blogger wrote in relation to Pink Lids, under the ironic heading "Let's start a Brown Colon Cancer Awareness campaign": "The latest in sexy pink breast cancer … comes from Mount Franklin, purveyors of pointless, wasteful, plastic-ridden, environmentally unfriendly bottled water" (Lauredhel 2008).

4. The majority of indigenous Australians are urban-dwelling. Only 10 percent of the country's indigenous population (517,200 in 2006, or 2.5 percent for the overall population) lives in the Northern Territory (Taylor and Bell 2001, 3).

5. Indigenous people are three times more likely to have diabetes than nonindigenous Australians; at the time of the NTER, the Northern Territory had the highest number of diabetes-related deaths per capita of population in the country. According to the Australian Human Rights and Equal Opportunities Commission, "Indigenous Australians do not currently have the same opportunity to be as healthy as the non-indigenous population" (Calma 2006).

6. See Foster (2008, 213–217), for an excellent analysis of this.

7. Thanks to Jennifer Biddle for making this point in conversation with the authors.

8. By mid-2013, the shift away from sugar to spring water and low- or no-kilojoule drinks was 4.2 percent (from CCA correspondence, October 16, 2013).

References

Abram, S., and M. E. Lien. 2011. Performing nature at world's ends. *Ethnos* 76 (1): 3–18.

Adolph, E. F. 1947. *Physiology of Man in the Desert*. New York: Interscience.

Agamben, G. 1998. *Homo Sacer: Sovereign Power and Bare Life*. Stanford, CA: Stanford University Press.

Agamben, G. 2005. *State of Exception*. Chicago: University of Chicago Press.

Alter, L. 2007. London food reviewer says no to bottled water. treehugger.com. http://www.treehugger.com/green-food/london-food-reviewer-says-no-to-bottled -water.html.

American Academy of Environmental Engineers and Scientists. 2014. 2014 Grand Prize. *Leadership and Excellence in Environmental Engineering and Science*. http://www .aaees.org/environmentalcommunicationsawards-winners-2014-grandprize.php.

Amin, A., and N. Thrift. 2013. *Arts of the Political: New Openings for the Left*. Durham, NC: Duke University Press.

Anderson, W. 2012. Asia as method in science and technology studies. *East Asian Science, Technology and Society* 6 (4): 445–451.

Angel, J. 2008. Bottled water or bottled environmental damage? The Greens Australia. http://archive.greens.org.au/node/874.

Arnold, E. 2006. Bottled water: Pouring resources down the drain? Earth Policy Institute, February 2. http://www.bvsde.paho.org/bvsacd/cd47/bottled.pdf.

Arvidsson, A. 2006. *Brands: Meaning and Value in Media Culture*. New York: Routledge.

Atkins, P. 2007. Laboratories, laws, and the career of a commodity. *Environment and Planning. D, Society & Space* 25:967–989.

Australian Bureau of Statistics. 2011. *Population by Age and Sex, Regions of Australia, 2011.* http://www.abs.gov.au/ausstats/abs@.nsf/Products/3235.0~2011~Main+Featur es~Northern+Territory.

Australian Financial Review 2008. Coca-Cola: Water, water everywhere ...: Case studies. http://www.afrbiz.com.au/case-studies/coca-cola-water-water-everywhere.html.

Bainbridge, J. 2009. Sector insight: Bottled water. *Marketing Magazine*, December 2. http://www.marketingmagazine.co.uk/article/971564/sector-insight-bottled-water.

Bakker, K. J. 2003. *An Uncooperative Commodity: Privatizing Water in England and Wales.* Oxford: Oxford University Press.

Bakker, K. J. 2007. The "commons" versus the "commodity": Alter-globalization, anti-privatization and the human right to water in the global south. *Antipode* 39:430–455.

Bakker, K. J. 2009. Neoliberal nature, ecological fixes, and the pitfalls of comparative research. *Environment & Planning A* 41:1781–1787.

Bakker, K. J. 2010. *Privatizing Water: Governance Failure and the World's Urban Water Crisis.* Ithaca, NY: Cornell University Press.

Barlow, M. 2007. *Blue Covenant: The Global Water Crisis and the Coming Battle for the Right to Water.* New York: New Press.

Barlow, M., and T. Clarke. 2002. *Blue Gold: The Fight to Stop the Corporate Theft of the World's Water.* New York: New Press.

Barnes, D. K. A., F. Galgani, R. C. Thompson, and M. Barlaz. 2009. Accumulation and fragmentation of plastic debris in global environments. *Philosophical Transactions of the Royal Society of London. Series B, Biological Sciences* 364 (1526): 1985–1998.

Barnett, M. 2008. *Bottled Water Coolers.* Jakarta: Asia Middle East Bottled Water Association.

Barry, A. 2004. Ethical capitalism. In *Global Governmentality*, ed. W. Larner and W. Walters, 195–211. London: Routledge.

Barry, A. 2005. Pharmaceutical matters. *Theory, Culture & Society* 22 (1): 51–69.

Barry, A. 2013. *Material Politics: Disputes along the Pipeline.* London: Wiley-Blackwell.

Barry, A., and D. Slater. 2002. Introduction: The technological economy. *Economy and Society* 31:175–193.

Battersby, L., and A. Holland. 2013. Thankyou water evangelist links revealed. *Sydney Morning Herald*, August 11. http://www.smh.com.au/national/thankyou -water-evangelist-links-revealed-20130810-2rp5e.html#ixzz2ib7wibYY.

BBC. 2004. Soft drink is purified tap water. *BBC News*, March 1.

BBC News South Asia. 2011. India Census: Population goes up to 1.21 bn. *BBC News South Asia,* March 31. http://www.bbc.co.uk/news/world-south-asia-12916888.

Bennett, J. 2001. *The Enchantment of Modern Life: Attachments, Crossings, and Ethics.* Princeton, NJ: Princeton University Press.

Bennett, J. 2004. The force of things: Steps toward an ecology of matter. *Political Theory* 32 (3): 347–372.

Bennett, J. 2007. Edible matter. *New Left Review* 45:133–145.

Bennett, J. 2010. *Vibrant Matter: A Political Ecology of Things.* Durham, NC: Duke University Press.

Bennett, T. 2009. Museum, field, colony: Colonial governmentality and the circulation of reference. *Journal of Cultural Economy* 2 (1–2): 99–116.

Bensaude-Vincent, B. 2013. Plastics, materials and dreams of dematerialization. In *Accumulation: The Material Politics of Plastic,* ed. J. Gabrys, G. Hawkins, and M. Michael, 17–29. Abingdon, UK: Routledge.

Bensaude-Vincent, B., and I. Stengers. 1996. *A History of Chemistry.* Cambridge, MA: Harvard University Press.

Berlant, L. G. 2011. *Cruel Optimism.* Durham, NC: Duke University Press.

Beverage Digest. 1999. Coke announces Dasani Water: CCE Schimberg cites appeal of purified water. *Beverage Digest,* February 19.

Bhushan, R. 2002. Bold and Bisleri. *The Hindu Business Line,* April 25. http://www .thehindubusinessline.in/catalyst/2002/04/25/stories/2002042500010100.htm.

Black, L. 2004. *Bottled Water.* YouTube. https://www.youtube.com/watch?v=NqX Frs6quvE.

Bottled Water Alliance. 2008. Bottled Water alliance website. http://www.bottled wateralliance.com.

Bourdieu, P. 1984. *A Social Critique of the Judgment of Taste.* Cambridge, MA: Harvard University Press.

Bowlby, R. 2000. *Carried Away: The Invention of Modern Shopping.* London: Faber and Faber.

Brandau, O. 2012. *Stretch Blow Molding.* Philadelphia: Elsevier.

Braun, B., and S. Whatmore, eds. 2010. *Political Matter: Technoscience, Democracy and Public Life.* Minneapolis: University of Minnesota Press.

Brewer, J., and F. Trentmann, eds. 2006. *Consuming Cultures: Global Perspectives, Historical Trajectories, Transnational Exchanges.* Oxford: Berg.

Brimblecombe, J., and D. Thomas. 2010. Income management isn't working. *Crikey*, May 17. http://www.crikey.com.au/2010/05/17/income-management-isnt-working -and-macklins-twisting-the-truth/?wpmp_switcher=mobile&wpmp_tp=1.

Briscoe, J., Malik, R. P. S., and World Bank. 2006. *India's Water Economy: Bracing for a Turbulent Future*. New Delhi: Oxford University Press.

Brita. 2008 FilterForGood: Filter for good. http://www.brita.com/filter-for-good.

Brita. 2010, *The Earth Needs Brita* [online video]. http://www.youtube.com/ watch?v=fjNfwyDBbJs.

Brooks, D., and G. Giles, eds. 2002. *PET Packaging Technology*. Sheffield: Sheffield Academic Press.

Brough, M. 2007. *National Emergency Response to Protect Aboriginal Children in the NT*. Media release. Canberra: Commonwealth Government.

Brown, B. 2003. *A Sense of Things: The Object Matter of American Literature*. Chicago: University of Chicago Press.

Butler, J. 2010. Performative agency. *Journal of Cultural Economy* 3 (2): 147–161.

Çalişkan, K., and M. Callon. 2009. Economization, part 1: Shifting attention from the economy towards processes of economization. *Economy and Society* 38 (3): 369–398.

Çalişkan, K., and M. Callon. 2010. Economization, part 2: A research programme for the study of markets. *Economy and Society* 39 (1): 1–32.

Callon, M., ed. 1998. *The Laws of the Markets*. Malden, MA: Blackwell.

Callon, M. 2008. An essay on framing and overflowing: Economic externalities revisited by sociology. In *The Laws of the Markets*, ed. M. Callon, 244–269. Oxford: Blackwell.

Callon, M. 2009. Civilizing markets: Carbon trading between in vitro and in vivo experiments. *Accounting, Organizations and Society* 34:535–548.

Callon, M., P. Lascoumes, and Y. Barthe. 2009. *Acting in an Uncertain World: An Essay on Technical Democracy*. Cambridge, MA: MIT Press.

Callon, M., C. Méadel, and V. Rabeharisoa. 2002. The economy of qualities. *Economy and Society* 31 (2): 194–218.

Callon, M., Y. Millo, and F. Muniesa, eds. 2007. *Market Devices*. Oxford: Blackwell.

Calma, T. 2006. Diabetes in indigenous communities. Speech. Australian Human Rights Commission: Society International Diabetes Forum. http://www.humanrights .gov.au/news/speeches/diabetes-indigenous-communities.

Campaign Brief. 2012. Mount Franklin extends its iconic Pink Lids campaign to help fund additional McGrath breast care nurses around Australia. *Campaign Brief,* September 18. http://www.campaignbrief.com/2012/09/mount-franklin-extends-its-ico .html.

Cantwell, T. 1980. *Run Australia Run! The First Book on Running and Fitness Written Specifically for Australia's Fun Runners.* Sydney: Horwitz.

CCA (Coca-Cola Amatil). 2009. Before and after—red to white activation. Unpublished Powerpoint Presentation. Coca-Cola Amatil Ltd.

CCA (Coca-Cola Amatil). 2010. Company history. http://ccamatil.com/ABOUTCCA/ Pages/CompanyHistory.aspx.

CCA (Coca-Cola Amatil). 2011. *Corporate Responsibility Report 2011: Case Studies.* http://cca2011crr.reportonline.com.au/case-studies.

CCA (Coca-Cola Amatil). 2012. *Fact Book 2012.* http://ccamatil.com/Investor Relations/ShareholderInfo/Documents/2012%20CCA%20Fact%20Book.pdf.

Chakrabarty, D. 2002. *Habitations of Modernity: Essays in the Wake of Subaltern Studies.* Chicago: University of Chicago Press.

Chatterjee, D. K. 2004. *The Ethics of Assistance: Morality and the Distant Needy.* New York: Cambridge University Press.

Clarke, T. 2007. *Inside the Bottle: An Exposé of the Bottled Water Industry.* Ottawa: Polaris Institute.

Clarke, T. 2010. Interview with Morgan Richards, February 19.

Cochoy, F. 1998. Another discipline for the market economy: Marketing as a performative knowledge and know-how for capitalism. In *The Laws of the Markets,* ed. M. Callon, 97–114. Malden, MA: Blackwell.

Cochoy, F. 2007. A sociology of market-things: On tending the garden of choices in mass retailing. In *Market Devices,* ed. M. Callon, Y. Millo, and F. Muniesa, 109–129. London: Blackwell.

Cochoy, F. 2012. Curiosity, packaging, and the economics of surprise. Charisma Consumer Market Studies. http://www.charisma-network.net/markets/curiosity -packaging.

Cochoy, F. 2013. Trojan cans. Limn, *Food Infrastructures.* http://limn.it/trojan-cans.

Cochoy, F., M. Giraudeau, and L. McFall. 2010. Performativity, economics and politics. *Journal of Cultural Economy* 3 (2): 139–146.

Cochoy, F., and C. Grandclément-Chaffy. 2005. Publicizing Goldilocks' choice at the supermarket: The political work of shopping packs, carts and talk. In *Making*

Things Public: Atmospheres of Democracy, ed. B. Latour and P. Weibel, 646–657. Cambridge, MA: ZKM and MIT Press.

Coley, N. G. 1990. Physicians, chemists, and the analysis of mineral waters: The most difficult part of chemistry. In *The Medical History of Waters and Spas*, ed. R. Porter, 56–66. London: Wellcome Institute for the History of Medicine.

Collier, S., and A. Lakoff. 2005. On regimes of living. In *Global Assemblages: Technology, Politics and Ethics as Anthropological Problems*, ed. A. Ong and S. Collier, 22–39. London: Blackwell.

Container Recycling Institute. 2003. *Report Shows Plastic Bottle Waste Tripled since 1995*. Media release, September 15. http://www.wvcag.org/news/fair_use/2003/09_16.htm.

Cooper, K. H. 1968. *Aerobics*. New York: M. Evans.

copyranter. 2008. So drinking bottled water is like giving my car a blow job? *copyranter*, May 15. http://copyranter.blogspot.com.au/2008/05/so-drinking-bottled-water-is-like.html.

Coren, G. 2008. Bling-bling guzzlers are now the new smokers. *The Times*, February 16.

Cormier, Z. 2008. Message in a Bottle. *New Internationalist* 415:10–12.

Corporate Accountability International. n.d. Think outside the bottle. http://www.stopcorporateabuse.org/campaigns/challenge-corporate-control-water/think-outside-bottle.

Costill, D. L. 1968. *What Research Tells the Coach about Distance Running*. Washington, DC: American Association for Health, Physical Education, and Recreation.

Crawford, R. 2006. Health as a meaningful docial practice. *Health* 10 (4): 401–420.

Cressy, J. 2010. Interview with Morgan Richards, May 27.

Cressy, J., Polaris Institute, and CUPE Nova Scotia. 2009. *Bottled Watergate: Why Is the Federal Government Spending Millions of Tax Dollars on Bottled Water?* Halifax: Polaris Institute and CUPE [Canadian Union of Public Employees] Nova Scotia.

Da Costa, B., and K. Philip. 2008. *Tactical Biopolitics: Art, Activism and Technoscience*. Cambridge, MA: MIT Press.

Dana, L. P. 2000. Evian water. *British Food Journal* 102 (5–6): 379–389.

Dasgupta, P. M. 2010. Water management. *Financial Express*, June 15.

Davies, W. 2013. When is a market not a market? "Exemption," "externality" and "exception" in the case of European state aid rules. *Theory, Culture & Society* 30 (2): 32–59.

Dean, J., J. W. Anderson, and G. Lovnik. 2006. *Reformatting Politics: Information Technology and Global Society*. New York: Routledge.

Deccan Chronicle. 2010a. Water lorries play truant in north Chennai areas. *Deccan Chronicle*, August 24. tniusnews.org/index.php?option=com_content&view=articl e&id=13442:water-lorries-play-truant-in-north-chennai-areas&catid=56:Water%20 Supply%20&Itemid=90.

Deccan Chronicle. 2010b. Water shortage cripples suburb. *Deccan Chronicle*, April 16. http://www.deccanchronicle.com/130403/news-current-affairs/article/drinking -water-crisis-suburbs.

DeLanda, M. 1995. Uniformity and variability: An essay in the philosophy of matter. Paper presented at Doors of Perception 3 Conference, Netherlands Design Institute, Amsterdam, November 7. http://www.museum.doorsofperception.com/doors3/ transcripts/Delanda.html.

DeLanda, M. 2006. *A New Philosophy of Society: Assemblage Theory and Social Complexity*. London: Continuum.

DeLanda, M. 2011. *Philosophy and Simulation: The Emergence of Synthetic Reason*. London: Continuum.

Deleuze, G. 1993. *The Logic of Sense*. Trans. M. Lester with C. Stivale. New York: Columbia University Press.

Deleuze, G., and F. Guattari. 2003. *A Thousand Plateaus*. Minneapolis: University of Minnesota Press.

Départment Médias Études et Communication Danone. 2006. *Building Danone: 30 Years of Passion*. Paris: Groupe Danone.

Desai, D. 2010. The looming water shortage. *Financial Chronicle*, March 8. http:// www.mydigitalfc.com/leisure-writing/looming-water-shortage-386.

Dieter, M. 2009. Processes, issues, air: Toward reticular politics. *Australian Humanities Review* 46:55–66.

DiGregorio, M. R. 1994. *Urban Harvest: Recycling as a Peasant Industry in Northern Vietnam*. Honolulu: East-West Center.

Do Something. 2008. *Do Something: Go Tap*. http://dosomething.net.au.

Drake, I. 2010. Asia boosts global bottled water market. *Australian Food News*, January 15. http://www.ausfoodnews.com.au/2010/01/15/asia-boosts-global-bottled-water -market.html.

Drummond, L. B. W. 2000. Street scenes: Practices of public and private space in urban Vietnam. *Urban Studies* 37 (12): 2377–2391.

Dumit, J. 2002. Drugs for life. *Molecular Interventions* 2 (3): 124–127.

Ellwood, W. 2008. This toxic life. *New Internationalist* 415:4–7.

Euromonitor International. 2010. *Drinking Cultures of the World: Globalisation Creates Opportunities.* http://www.euromonitor.com.

Evian. 2011. Interview with PR representative, July.

Evian. 2014. At home and on the go. Advertisement. http://www.evian.com/en_INT/54-evian-PET-bottles.

Feher, M. 2007. *Nongovernmental Politics.* London: Zone Books.

Fixx, J. F. 1977. *The Complete Book of Running.* Melbourne: Outback Press.

Foster, R. J. 2008. *Coca-Globalization: Following Soft Drinks from New York to New Guinea.* New York: Palgrave Macmillan.

Foucault, M. 2008. *The Birth of Biopolitics: Lectures at the Collège de France 1978–79.* Trans. G. Burchell. Basingstoke: Palgrave Macmillan.

Francis, C. 2009. *Inquiry into Community Stores in Remote Aboriginal and Torres Strait Islander Communities. Report to House of Representatives.* Canberra: Commonwealth Government.

Frederick, P. 2003. Chennai's water woes. *The Hindu,* June 5. http://www.hindu.com/thehindu/mp/2003/06/05/stories/2003060500290100.htm.

Freinkel, S. 2011. *Plastic: A Toxic Love Story.* Melbourne: Text.

Frow, J. 1995. *Cultural Studies and Cultural Value.* Oxford: Oxford University Press.

Frow, J. 2002. Signature and brand. In *High-Pop: Making Culture into Popular Entertainment,* ed. J. Collins, 56–74. Oxford: Blackwell.

Fry, C. 2007. Springs and roundabouts. *The Guardian,* March 23.

Gabrys, J. 2011. *Digital Rubbish: A Natural History of Electronics.* Ann Arbor: University of Michigan Press.

Gabrys, J. 2013. Plastic and the work of the biodegradable. In *Accumulation: The Material Politics of Plastic,* ed. J. Gabrys, G. Hawkins, and M. Michael, 208–227. Abingdon, UK: Routledge.

Gandy, M. 1994. *Recycling and the Politics of Urban Waste.* New York: St. Martin's Press.

Gandy, M. 2006. Zones of indistinction: Bio-political contestations in the urban arena. *Cultural Geographies* 13:497–516.

Gandy, M. 2008. Landscapes of disaster: Water, modernity, and urban fragmentation. *Environment & Planning A* 40 (1): 108–130.

Garrett, B. 2004. Coke's water bomb. *BBC Money Programme,* June 16.

Gibson, J. 2007. Coke binge is the real unhealthy thing in Territory, where sugar's a vice. *Sydney Morning Herald*, September 18.

Gibson-Graham, J. K. 2006. *A Postcapitalist Politics*. Minneapolis: University of Minnesota Press.

Giles, G., and G. Bockner. 2002. Commercial considerations. In *PET Packaging Technology*, ed. D. Brooks and G. Giles, 26–35. Sheffield, UK: Sheffield Academic Press.

Gill, K. 2010. *Of Poverty and Plastic: Scavenging and Scrap Trading Entrepreneurs in India's Urban Informal Economy*. New Delhi: Oxford University Press.

Glycemic Index Foundation 2009. Submission into community stores in Remote Aboriginal and Torres Strait Islander Communities. Inquiry into Community Stores in Remote Aboriginal and Torres Strait Islander Communities. Canberra: Commonwealth of Australia.

Godden, C. 2011. Waste pickers in Asia: Contesting value and values. In *Inter-Asia Roundtable 2011: Recycling Cities*, ed. T. Bunnell, M. Miller, P. Marolt, L. Hongyan, and V. Yeo, n.p. Singapore: Asia Research Institute, National University of Singapore.

Gopakumar, G. 2012. *Transforming Urban Water Supplies in India: The Role of Reform and Partnerships in Globalization*. London: Routledge.

Gosford, B. 2014. Outback stores takes a dump on confidential Centrelink data at Yuendemu. *Crikey*, April 30. http://blogs.crikey.com.au/northern/2009/04/30/outback-stores-takes-a-dump-on-confidential-centrelink-data-at-yuendumu.

Grant, R. 2009. Message in a bottle. *Sydney Morning Herald Good Weekend*, June 20.

Green, M., and T. Green. 1985. *The Good Water Guide: The World's Best Bottled Waters*. London: Rosendale Press.

Gregson, N., H. Watkins, and M. Calestani. 2013. Political markets: Recycling, economization and marketization. *Economy and Society* 42:1–25.

Gruden 2008. Gruden launches Bottled Water Alliance. http://blog.gruden.com/2008/12/gruden-launches-bottled-water-alliance.

Grundy, J. 2008. Mount Franklin—greenwashing or not? Go greener, Australia: You know you want to. *Go Greener, Australia*, February 18, http://pandora.nla.gov.au/pan/81783/20080403-1320/www.gogreeneraustralia.com/blog/index.php/2008/02/index.html.

Harman, G. 2009. *Prince of Networks*. Melbourne: re.press.

Hawkins, G. 2011. Packaging water: Plastic bottles as market and public devices. *Economy and Society* 40 (4): 534–553.

Hein, K. 2002. Fashion conscious Dasani makes a splash at N.Y. design school. *Brandweek* 43 (38): 35.

Heynen, N., M. Kaika, and E. Swyngedouw. 2013. *In the Nature of Cities: Urban Political Ecology and the Politics of Urban Metabolism*. Hoboken, NJ: Taylor and Francis.

Hickman, L. 2009. Why bottled bling H_2O is an eco low. *The Guardian* environment blog, March 20. http://www.theguardian.com/environment/blog/2009/mar/20/bling-h2o-bottled-water.

Hine, T. 1995. *The Total Package: The Secret History and Hidden Meanings of Boxes, Bottles, Cans, and Other Persuasive Containers*. Canada: Little, Brown.

Honig, B., and J. Gelonesi. 2013. The importance of public things. *The Philosopher's Zone*. ABC Radio National, April 21. http://www.abc.net.au/radionational/programs/philosopherszone/honig/4623818.

Howard, T. 1999a. Coke aims below-the-line on Dasani. *Brandweek* 40 (11): 14.

Howard, T. 1999b. Coke ads peg Desani as a path to serenity. *Brandweek* 40 (32): 12.

Hydration for Health Initiative. 2014. Website. http://www.h4hinitiative.com.

International Bottled Water Association. 2014. Website. http://www.bottledwater matters.org/taxonomy/term/International%20Bottled%20Water%20Association.

Ivester, D. 1998. Coke nears the water's edge. Named "Dasani"? New package. Impact on other waters in Coke system? *Beverage Digest*, June 11.

Jacobson, M. 2005. *Liquid Candy: How Soft Drinks Are Harming Americans' Health*. Washington, DC: Center for Science in the Public Interest.

Janakarajan, S., J. Butterworth, P. Moriarty, and C. Batchelor. 2007. Strengthened city, marginalised peri-urban villages: Stakeholder dialogues for inclusive urbanisation in Chennai, India. In *Peri-Urban Water Conflicts: Supporting Dialogue and Negotiation*, ed. J. Butterworth et al., n.p. Delft: IRC International Water and Sanitation Centre.

Judge, M. 2006. *Idiocracy*. Twentieth Century Fox.

Kaika, M. 2006. The political ecology of water scarcity: The 1989–1991 Athenian drought. In *In the Nature of Cities: Urban Political Ecology and the Politics of Urban Metabolism*, ed. N. Heynen, M. Kaika, and E. Swyngedouw, 150–164. London: Routledge.

Kamat, A. 2002. *Water Profiteers*. New Delhi: India Resource Centre. http://www.indiaresource.org/issues/water/2003/waterprofiteers.html.

Kjellén, K., and G. McGranahan. 2006. *Informal Water Vendors and the Urban Poor*. London: International Institute for Environment and Development.

Knorr Cetina, K. 1997. Sociality with objects: Social relations in postsocial knowledge societies. *Theory, Culture & Society* 14 (4): 1–30.

Kolata, G. 2001. Scientist at work: David Costill, a career spent in study of training and exercise, lap by grueling lap. *New York Times*, October 30.

Lahiri-Dutt, K. 2008. Guest editorial: The quest for water: Rethinking water scarcity. *Development* 51:5–11.

Lakshmi, K. 2005. Chennai: A city in deep water. *The Hindu*, April 16. http://www .hindu.com/pp/2005/04/16/stories/2005041600040100.htm.

Lakshmi, K. 2009. Decline in groundwater levels. *The Hindu*, October 31. http:// www.hindu.com/pp/2009/10/31/stories/2009103150020100.htm.

Landry, C. J. 2002. Overview: The new economy of water. *Water Resources Impact* 4 (1): 2–3.

Latour, B. 1999. *Pandora's Hope: Essays on the Reality of Science Studies*. Cambridge, MA: Harvard University Press.

Latour, B. 2004. How to talk about the body: The normative dimension of science studies. *Body & Society* 10 (2–3): 205–229.

Latour, B. 2005. From realpolitik to dingpolitik or how to make things public. In *Making Things Public: Atmospheres of Democracy*, ed. B. Latour and P. Weibel, 14–43. Cambridge, MA: MIT Press.

Lauredhel. 2008. Mount Franklin breast cancer ads. Let's start a brown colon cancer awareness campaign. http://viv.id.au/blog/20081011.2306/quick-hit-mount-franklin -breast-cancer-ads.

Law, J. 2004a. *After Method: Mess in Social Science Research*. London: Routledge.

Law, J. 2004b. And if the global were small and noncoherent? Method, complexity, and the Baroque. *Environment and Planning D, Society & Space* 22:13–26.

Law, J. 2004c. Matter-ing: Or how might STS contribute? http://www.lancaster.ac .uk/sociology/research/publications/papers/law-matter-ing.pdf.

Law, J. 2009. Collateral realities. http://heterogeneities.net/publications/Law2009 CollateralRealities.pdf.

Lawrence, F. 2004a. Soft drink is purified tap water. *BBC News*, March 1. http://news .bbc.co.uk/2/hi/uk_news/3523303.stm

Lawrence, F. 2004b. Things get worse with Coke: Bottled tap water withdrawn after cancer scare. *The Guardian*, March 20.

Le Guen, V. n.d. Right packaging product. http://www.sidel.com/about-sidel/inline -magazine/right-packaging-product.

Lee, J. 2008. Message on a bottle labelled as greenwash. *Sydney Morning Herald*, February 23.

Lipperts, B. 2003. Muddy water. *Adweek* 44 (27): 22.

Lury, C. 2004. *Brands: The Logos of the Global Economy*. New York: Routledge.

Lury, C. 2009. Brand as assemblage. *Journal of Cultural Economy* 21 (1–2): 67–82.

MacBride, S. 2011. *Recycling Reconsidered: The Present Failure and Future Promise of Environmental Action in the United States*. Cambridge, MA: MIT Press.

Mackaman, D. P. 1998. *Leisure Settings: Bourgeois Culture, Medicine, and the Spa in Modern France*. Chicago: University of Chicago Press.

Manly Council. 2009. Do something! http://www.google.com.au/search?client=safa ri&rls=en&q=manly+council+water+fountain&ie=UTF-8&oe=UTF-8&gfe_rd=cr&ei= 1gx9U5u1NsqN8QfK4YGwBQ.

Manly Council. 2010. Manly Council's filtered water bubbler project. http://www .manly.nsw.gov.au/IgnitionSuite/uploads/docs/Filtered%20Bubbler%20Project(1) .pdf.

Mariappan, J. 2011. City sees jump in sale of water cans. *Times of India*, April 12. http://articles.timesofindia.indiatimes.com/2011-04-12/chennai/29409770_1 _drinking-water-water-supply-mld.

Marres, N. 2007. The issues deserve more credit: Pragmatist contributions to the study of public involvement in controversy. *Social Studies of Science* 37 (5): 759–781.

Marres, N. 2012. *Material Participation: Technology, the Environment and Everyday Publics*. Basingstoke, UK: Palgrave Macmillan.

Marres, N., and R. Rogers. 2005. Recipe for tracing the fate of issues and their publics on the web. In *Making Things Public: Atmospheres of Democracy*, ed. B. Latour and P. Weibel, 922–935. Cambridge, MA: MIT Press.

Marty, N. 2006. La consommation des eaux embouteillées. *Vingtième Siècle. Revue d'histoire* 91 (3): 25–41.

McCarthy, D. 2009. BBC accused of wasting £406,000 of public money a year on bottled water. *The Guardian*, August 11.

McCartney, M. 2011. Waterlogged? The facts behind the claims that we all need to drink more water (life). *Student BMJ* 19:12–14.

McFall, L. 2009. Devices and desires: How useful is the "new" economic sociology for understanding market attachment? *Sociology Compass* 3 (2): 267–282.

Mehra, R., et al. 1996. Women in waste collection and recycling in Hochiminh city. *Population and Environment* 18 (2): 187–199.

Mehta, L. 2007. Whose scarcity? Whose property? The case of water in western India. *Land Use Policy* 24:654–663.

Meikle, J. 1995. *American Plastic: A Cultural History.* New Brunswick, NJ: Rutgers University Press.

Merrett, N. 2008. "Ethical" water brands may boost flagging UK sales. *Beverage Daily,* April 8. http://www.beveragedaily.com/Industry-Markets/Ethical-water-brands-may-boost-flagging-UK-sales.

Michael, M. 2000. *Reconnecting Culture, Technology, and Nature: From Society to Heterogeneity.* London: Routledge.

Michael, M. 2002. Comprehension, apprehension, prehension: Heterogeneity and the public understanding of science. *Science, Technology & Human Values* 27 (3): 357–378.

Michael, M., and M. Rosengarten. 2012. HIV, globalization and topology: Of prepositions and propositions. *Theory, Culture & Society* 29 (4–5): 1–23.

Miller, P. 2004. Lewis Black: On Broadway. HBO, May 15.

Miller, P., and N. Rose. 2008. *Governing the Present: Administering Economic, Social and Personal Life.* Cambridge, MA: Polity Press.

Mitchell, C. L. 2009. Trading trash in the transition: Economic restructuring, urban spatial transformation, and the boom and bust of Hanoi's informal waste trade. *Environment & Planning A* 41:2633–2650.

Mol, A. 2002. *Body Multiple: Ontology in Medical Practice.* Durham, NC: Duke University Press.

Mol, A. 2008. *The Logic of Care: Health and the Problem of Patient Choice.* London: Routledge.

Moor, L. 2007. *The Rise of Brands.* Oxford: Berg.

Morgan, B. 2006. Emerging global water welfarism: Access to water, unruly consumers and transitional government. In *Consuming Cultures: Global Perspectives, Historical Trajectories, Transnational Exchanges,* ed. J. Brewer and F. Trentmann, 279–303. Oxford: Berg.

Moynihan, R. 2003. The making of a disease: Female sexual dysfunction. *British Medical Journal* 27 (3): 357–378.

Moynihan, R., and D. Henry. 2006. The fight against disease mongering: Generating knowledge for action. *PLoS Medicine* 3 (4): 1–4.

Muernmart, A. 2008. *The "War" of Waters: Nestlé vs Singha.* Jakarta: Asia Middle East Bottled Water Association.

Muniesa, F. 2012. A flank movement in the understanding of valuation. *Sociological Review* 59 (2): 24–38.

Muniesa, F., Y. Millo, and M. Callon. 2007. An introduction to market devices. In *Market Devices*, ed. M. Callon, Y. Millo, and F. Muniesa, 1–12. Oxford: Blackwell.

Murthy, L. 2005. *Boond-boond mein paisa*: Bottled water is big business. *Infochange Agenda*. http://infochangeindia.org/agenda/the-politics-of-water/boond-boond-mein -paisa-bottled-water-is-big-business.html.

Nadu, T. 2007. Nine water packaging units closed. *The Hindu*, August 13. http:// www.hindu.com/2007/08/13/stories/2007081351950300.htm.

NAPCOR (National Association for PET Container Resources) n.d. PET basics. http:// www.napcor.com.

Narain, B. L. 2005. Water scarcity in Chennai, India. Centre for Civil Society. http:// ccs.in/internship_papers/2005/7.%20Water%20scarcity%20in%20Chennai.pdf.

Northern Territory Emergency Response Taskforce. 2008. *Final Report to Government, June 2008*. Canberra: Commonwealth Government.

Novas, C., and N. Rose. 2005. Biological citizenship. In *Global Assemblages: Technology, Politics, and Ethics as Anthropological Problems*, ed. A. Ong and S. J. Collier. Malden, MA: Blackwell.

Olson, E. D., Poling, D., Solomon, G., and Natural Resources Defense Council. 1999. *Bottled Water: Pure Drink or Pure Hype?* Washington, DC: Natural Resources Defense Council.

One Water Australia. 2013. Do one good thing: Our work in Australia. http://www .doonegoodthing.com.au/our-work/our-work-in-australia.

Ong, A. 2006. Mutations in citizenship. *Theory, Culture & Society* 23 (2–3): 499–505.

Ong, A., and S. J. Collier, eds. 2005. *Global Assemblages: Technology, Politics, and Ethics as Anthropological Problems*. Malden, MA: Blackwell.

Palmer, D. 2009. Ethical water brand to donate all profits to clean water projects. *Australian Food News*, January 14. http://www.ausfoodnews.com.au/ 2009/01/14/ ethical-water-brand-to-donate-all-profits-to-clean-water-projects.html.

Pearse, W. 2010. A look at Vietnam's plastic craft villages. *Our World*. http:// ourworld.unu.edu/en/a-look-at-vietnam%E2%80%99s-plastic-craft-villages.

PET remains a favourite in the beverage industry. 2008. *Drink Technology and Marketing*, March.

Pignarre, P., and I. Stengers. 2011. *Capitalist Sorcery: Breaking the Spell*. Basingstoke: Palgrave Macmillan.

Plastics Academy. n.d. Hall of Fame. http://www.plasticshalloffame.com.

Polaris Institute. 2007. Inside the Bottle: The people's campaign on the bottled water industry. http://www.insidethebottle.org/Home.html.

Polaris Institute. 2009. *Murky Waters: The Urgent Need for Health and Environmental Regulations of the Bottled Water Industry.* Ottawa: Polaris Institute.

Polaris Institute. 2014. About us. http://www.polarisinstitute.org/aboutus.

Porter, R., ed. 1990. *The Medical History of Waters and Spas.* London: Wellcome Institute for the History of Medicine.

Potter, E. 2009a. Interview with S. Raghavan, Chennai, June 12.

Potter, E. 2009b. Interview with C. Kurian, Chennai, June 12.

Potter, E. 2009c. Interview with C. Kurian. Chennai, June 13.

Potter, E. 2009d. Interview with CAA spokesperson, CAA headquarters, July 3.

Potter, E. 2009e. Interview with CAA spokesperson, CAA headquarters, December 10.

Potter, E. 2011. Interview with CAA spokesperson, CAA headquarters, November 2.

Prescott, S. 2012. Thankyou water: A social enterprise making waves. *Mamamia.* http://www.mamamia.com.au/mamamia-cares/thankyou-water-a-social-enterprise-making-waves.

Rabinow, P., and N. Rose. 2006. Biopower today. *Biosocieties* 1 (2): 195–217.

Race, K. 2009. *Pleasure Consuming Medicine: The Queer Politics of Drugs.* Durham, NC: Duke University Press.

Raja, M. 2008. Public water, privately bottled profits. *Asia Times,* May 8. http://atimes.com/atimes/South_Asia/JE08Df04.html.

Rani, B., R. Maheshwari, A. Garg, and M. Prasad. 2012. Bottled water: A global market overview. *Bulletin of Environment, Pharmacology and Life Sciences* 1 (6): 1–4.

Raturi, P. 2005. And this is how Parle Bisleri began. redif.com, June 10. http://www.rediff.com/money/2005/jun/10spec.htm.

Ray, B. 2008. *Water: The Looming Crisis in India.* Lanham, MD: Lexington Books.

Renuka, M. 2001. Liquid asset. *India Today,* May 14. http://indiatoday.intoday.in/story/multi-crore-water-industry-attracts-many-players-purity-to-decide-who-stays-in-business/1/233734.html.

Rico, H. 2011. Indian bottled water market expecting rapid growth. *PRNewswire.* http://www.prnewswire.com/news-releases/indian-bottled-water-market-expecting-rapid-growth-118578544.html.

Robinson, I. 2007. The consumer dimension of stakeholder activism: The anti-sweatshop movement in the United States. In *Non-Governmental Politics,* ed. M. Feher, Y. McKee, and G. Krikorian, 200–222. London: Zone Books.

Rodwan, J.G.J. 2011. *Bottled Water 2011: The Recovery Continues*. International Bottled Water Association Market Report. http://www.bottledwater.org/files/2011 BWstats.pdf.

Rose, N. 1998. *Inventing Our Selves: Psychology, Power and Personhood*. Cambridge: Cambridge University Press.

Royte, E. 2008. *Bottlemania: How Water Went on Sale and Why We Bought It*. New York: Bloomsbury.

Rucknel, C. 2006. *The Vietnamese Plastics Industry: Rapid Growth and Much Future Promise*. http://www.business-in-asia.com/plastics_in_vietnam.html.

Schatzki, T. R. 1996. *Social Practices: A Wittgensteinian Approach to Human Activity and the Social*. Cambridge: Cambridge University Press.

Schatzki, T. 2009. Timespace and the organization of social life. In *Time, Consumption and Everyday Life: Practice, Materiality and Culture*, ed. E. Shove, F. Trentmann, and R. Wilk, 35–48. Oxford: Berg.

Shiva, V. 2002. *Water Wars: Privatisation, Pollution and Profit*. Cambridge, MA: South End Press.

Shove, E. 2003. *Comfort, Cleanliness and Convenience: The Social Organization of Normality*. Oxford: Berg.

Shove, E., M. Pantzar, and M. Watson. 2012. *The Dynamics of Social Practice: Everyday Life and How It Changes*. London: Sage.

Shove, E., F. Trentmann, and R. Wilk. 2009. *Time, Consumption and Everyday Life: Practice, Materiality and Culture*. Oxford: Berg.

Shove, E., M. Watson, and M. Hand, eds. 2007. *The Design of Everyday Life*. Oxford: Berg.

Skelton, R. 2008. In the outback, a Third World utopia. *Sydney Morning Herald*, March 1.

Smith, B. H. 1988. *Contingencies of Value: Alternative Perspectives for Critical Theory*. Cambridge, MA: Harvard University Press.

Sofia, Z. 2000. Container technologies. *Hypatia* 15 (2): 181–201.

Stengers, I. 2008. A constructivist reading of *Process and Reality*. *Theory, Culture & Society* 25 (4): 91–110.

Story of Stuff Project. 2010. The story of bottled water. http://www.storyofstuff.org/ movies-all/story-of-bottled-water.

Sujatha, R. 2009. Contaminated water causes diseases, allege residents. *The Hindu*, June 12. http://www.thehindu.com/todays-paper/tp-national/tp-tamilnadu/conta minated-water-causes-diseases-allege-residents/article256146.ece.

Take Back the Filter. 2008. Take back the filter. http://www.takebackthefilter.org.

Taylor, A., and L. Bell. 2001. *The Northern Territory's Declining Share of Australia's Indigenous Population: A Call for a Research Agenda*. Darwin, Australia: Northern Institute, Charles Darwin University.

Thankyou Water. 2008. We exist to fund life-changing water projects in developing nations. http://thankyou.co/water.

Thirsty Planet Ltd. 2007. Website. http://www.thirsty-planet.com.

Thomas, N. 1991. *Entangled Objects: Exchange, Material Culture and Colonialism in the Pacific*. Cambridge, MA: Harvard University Press.

Times of India. 2010. Metrowater to launch SMS complaint facility. October 27. http://articles.timesofindia.indiatimes.com/2010-10-27/chennai/28228598_1 _metrowater-complaint-depots.

Timm, J. C. 2014. San Francisco bans sale of plastic water bottles on city property. MSNBC, March 13. http://www.msnbc.com/msnbc/san-francisco-bans-sale-plastic -water-bottles-climate-change.

Twede, D. 2012. The birth of modern packaging: Cartons, cans and bottles. *Journal of Historical Research in Marketing* 4 (2): 245–272.

URENCO and ENTEC 2009. Study of empty bottles in Hanoi. Manuscript. Hanoi: URENCO and ENTEC.

Vaidyanathan, A., and J. Saravanan. 2001. *Managing Water in Chennai*. New Delhi: Centre for Science and Environment.

Valtin, H. 2002. "Drink at least eight glasses of water a day." Really? Is there scientific evidence for "8 × 8"? *American Journal of Physiology* 283(5): R993–R1004.

van Tubergen, A., and S. van der Linden. 2002. A brief history of spa therapy. *Annals of the Rheumatic Diseases* 61:273–275.

Varadarajan, R., and A. Menon. 1988. Cause-related marketing: A coalignment of marketing strategy and corporate philanthropy. *Journal of Marketing* 52 (July): 58–74.

Varma, M. D. 2009. The changing profile of "missed call" makers. *The Hindu*, June 11. http://www.hindu.com/2009/06/11/stories/2009061157720200.htm.

Velloor, R. 2005. Water's the next big business in India. *Asia One*, May 31. http:// eresources.nlb.gov.sg/newspapers/Digitised/Article/straitstimes20050531-1.2.5.aspx.

Waldby, C. 2000. *The Visible Human Project: Informatic Bodies and Posthuman Medicine*. London: Routledge.

Wanctin, L., K. Dalmeny, J. Longfield, and Sustain. 2006. *Have You Bottled It? How Drinking Tap Water Can Help Save You and the Planet*. London: Sustain.

Ward, L., O. Cain, R. Mullally, K. Holliday, A. Wernham, P. Baillie, and S. Greenfield. 2009. Health beliefs about bottled water: A qualitative study. *BMC Public Health* 9 (1): 196–210.

Warner, M. 2002. *Publics and Counterpublics*. London: Zone Books.

Waste & Recycling Action Programme (WRAP). 2006. UK Plastic Bottle Recycling Survey 2006, March. http://www.wrap.org.uk/applications/publications/publication _details.rm?id=698&publication=2768.

Whatmore, S. J., and C. Landström. 2011. Flood apprentices: An exercise in making things public. *Economy and Society* 40 (4): 582–610.

Whitehead, A. N. 1978. *Process and Reality: An Essay in Cosmology*. New York: The Free Press.

Wilk, R. 2006. Bottled water: The pure commodity in the age of branding. *Journal of Consumer Culture* 6 (3): 303–325.

World Water Council. 1996. Website. http://www.worldwatercouncil.org.

Index